Springer Tracts in Modern Physics
Volume 176

Managing Editor: G. Höhler, Karlsruhe

Editors: H. Fukuyama, Kashiwa
J. Kühn, Karlsruhe
Th. Müller, Karlsruhe
A. Ruckenstein, New Jersey
F. Steiner, Ulm
J. Trümper, Garching
P. Wölfle, Karlsruhe

Honorary Editor: E. A. Niekisch, Jülich

Starting with Volume 165, Springer Tracts in Modern Physics is part of the Springer LINK service. For all customers with standing orders for Springer Tracts in Modern Physics we offer the full text in electronic form via LINK free of charge. Please contact your librarian who can receive a password for free access to the full articles by registration at:

http://link.springer.de/series/stmp/reg_form.htm

If you do not have a standing order you can nevertheless browse through the table of contents of the volumes and the abstracts of each article at:

http://link.springer.de/series/stmp/

There you will also find more information about the series.

Springer-Verlag Berlin Heidelberg GmbH

Physics and Astronomy ONLINE LIBRARY

http://www.springer.de/phys/

Springer Tracts in Modern Physics

Springer Tracts in Modern Physics provides comprehensive and critical reviews of topics of current interest in physics. The following fields are emphasized: elementary particle physics, solid-state physics, complex systems, and fundamental astrophysics.

Suitable reviews of other fields can also be accepted. The editors encourage prospective authors to correspond with them in advance of submitting an article. For reviews of topics belonging to the above mentioned fields, they should address the responsible editor, otherwise the managing editor.
See also http://www.springer.de/phys/books/stmp.html

Managing Editor

Gerhard Höhler

Institut für Theoretische Teilchenphysik
Universität Karlsruhe
Postfach 69 80
76128 Karlsruhe, Germany
Phone: +49 (7 21) 6 08 33 75
Fax: +49 (7 21) 37 07 26
Email: gerhard.hoehler@physik.uni-karlsruhe.de
http://www-ttp.physik.uni-karlsruhe.de/

Elementary Particle Physics, Editors

Johann H. Kühn

Institut für Theoretische Teilchenphysik
Universität Karlsruhe
Postfach 69 80
76128 Karlsruhe, Germany
Phone: +49 (7 21) 6 08 33 72
Fax: +49 (7 21) 37 07 26
Email: johann.kuehn@physik.uni-karlsruhe.de
http://www-ttp.physik.uni-karlsruhe.de/~jk

Thomas Müller

Institut für Experimentelle Kernphysik
Fakultät für Physik
Universität Karlsruhe
Postfach 69 80
76128 Karlsruhe, Germany
Phone: +49 (7 21) 6 08 35 24
Fax: +49 (7 21) 6 07 26 21
Email: thomas.muller@physik.uni-karlsruhe.de
http://www-ekp.physik.uni-karlsruhe.de

Fundamental Astrophysics, Editor

Joachim Trümper

Max-Planck-Institut für Extraterrestrische Physik
Postfach 16 03
85740 Garching, Germany
Phone: +49 (89) 32 99 35 59
Fax: +49 (89) 32 99 35 69
Email: jtrumper@mpe-garching.mpg.de
http://www.mpe-garching.mpg.de/index.html

Solid-State Physics, Editors

Hidetoshi Fukuyama
Editor for The Pacific Rim

University of Tokyo
Institute for Solid State Physics
5-1-5 Kashiwanoha, Kashiwa-shi
Chiba-ken 277-8581, Japan
Phone: +81 (471) 36 3201
Fax: +81 (471) 36 3217
Email: fukuyama@issp.u-tokyo.ac.jp
http://www.issp.u-tokyo.ac.jp/index_e.html

Andrei Ruckenstein
Editor for The Americas

Department of Physics and Astronomy
Rutgers, The State University of New Jersey
136 Frelinghuysen Road
Piscataway, NJ 08854-8019, USA
Phone: +1 (732) 445 43 29
Fax: +1 (732) 445-43 43
Email: andreir@physics.rutgers.edu
http://www.physics.rutgers.edu/people/pips/
Ruckenstein.html

Peter Wölfle

Institut für Theorie der Kondensierten Materie
Universität Karlsruhe
Postfach 69 80
76128 Karlsruhe, Germany
Phone: +49 (7 21) 6 08 35 90
Fax: +49 (7 21) 69 81 50
Email: woelfle@tkm.physik.uni-karlsruhe.de
http://www-tkm.physik.uni-karlsruhe.de

Complex Systems, Editor

Frank Steiner

Abteilung Theoretische Physik
Universität Ulm
Albert-Einstein-Allee 11
89069 Ulm, Germany
Phone: +49 (7 31) 5 02 29 10
Fax: +49 (7 31) 5 02 29 24
Email: steiner@physik.uni-ulm.de
http://www.physik.uni-ulm.de/theo/theophys.html

Martina Havenith

Infrared Spectroscopy of Molecular Clusters

An Introduction to Intermolecular Forces

With 34 Figures

Springer

Professor Martina Havenith

University of Bochum
Physical Chemistry II
Universitätsstrasse 150
44780 Bochum, Germany
E-mail: Martina.Havenith@ruhr-uni-bochum.de

Library of Congress Cataloging-in-Publication Data.

Die Deutsche Bibliothek - CIP-Einheitsaufnahme

Havenith, Martina:
Infrared spectroscopy of molecular clusters: an introduction to
intermolecular forces/Martina Havenith. – Berlin; Heidelberg;
New York; Barcelona; Hong Kong; London; Milan; Paris; Tokyo:
Springer, 2002
(Springer tracts in modern physics; Vol. 176)
(Physics and astronomy online library)

Physics and Astronomy Classification Scheme (PACS): 33.30 Ea and 36.40 Mr

ISSN print edition: 0081-3869
ISSN electronic edition: 1615-0430

ISBN 978-3-662-14648-4   ISBN 978-3-540-45457-1(eBook)
DOI 10.1007/978-3-540-45457-1

http://www.springer.de

© Springer-Verlag Berlin Heidelberg 2002
Originally published Springer-Verlag Berlin Heidelberg New York in 2002
Softcover reprint of the hardcover 1st edition 2002

Typesetting: Camera-ready copy from the author using a Springer LaTeX macro package
Cover design: *design & production* GmbH, Heidelberg

Printed on acid-free paper   SPIN: 10697257   56/3141/tr   5 4 3 2 1 0

Preface

This book is intended to give an introduction to intermolecular forces from an experimental point of view. Within the last 10 years the interest has turned more and more into an understanding of the weak, but important, intermolecular forces. New experimental techniques have been developed which have helped to gain more insight into this interesting topic.

This book is intended as an introduction for graduate students who are familiar with the main concepts of molecular spectroscopy. Special emphasis will be laid on the theoretical concepts.

After a detailed description of experimental techniques, the results for two prototype systems which have been the subject of several studies in the literature within recent years will be presented.

Ar–CO is becoming the most extensively studied van der Waals complex, theoretically and experimentally. Nevertheless, this example shows that even though the theory has greatly improved and has helped us to improve our knowledge of intermolecular forces, even for relatively simple cases the theory can still fall short of an accurate description.

For a long time $(NH_3)_2$ was considered as a prototype for hydrogen bonding. However, subsequent experimental and theoretical studies have revealed the mysteries of the obtained spectra and proved that our previous concept of hydrogen bonds was just too naive.

It is obvious that we need the best and most sensitive experimental techniques in order to provide the theory with benchmark systems. As long as we are not able to understand these prototype systems in detail it is hard to believe that we can really understand and predict hydrogen bonding in biologically relevant systems.

This book is based on my Habilitation work in Bonn. As such it has benefited from the work of a lot of people who have helped me over a long period of time. I specially want to mention the former students M. Petri, G. Hilpert, St. König, M. Scherer, I. Scheele, R. Lehnig, U. Merker, P. Engels and F. Madeja, who all contributed to this work. I want to thank Prof. Urban, who introduced me to IR spectroscopy and has been a great teacher and a wonderful person. As a second adviser on this work I want to thank Prof. G. Scoles, who was in Bonn on a Humboldt fellowship, and contributed with many stimulating discussions and shared his great enthusiasm for this topic.

The Sonderforschungsbereich SFB 334 was the basis on which this work evolved. In this context I wish to thank the DFG for their financial support and the theoreticians Prof. B. Hess and Dr. G. Jansen for their scientific contribution, especially on the Ar–CO project. The work on the ammonia dimer is based on a cooperation with several institutes and people. Most of the work was performed in Nijmegen with Dr. W.L. Meerts and Dr. H. Linnartz. Prof. Ad van der Avoird finally solved the riddle of the ammonia dimer. Furthermore, I want to thank W. Stahl and U. Buck for sharing their data with me.

I especially thank Roger Miller for his helpful comments on the manuscript. I want to thank Springer-Verlag for its patience and J. Masuch for editing the final version of this book.

This book is dedicated to my husband Albert Newen and my mother Eva Wittmann.

Bochum, November 2001   *M. Havenith*

Contents

1. Introduction

1.1 Motivation

The subject of the book is intermolecular forces. Intermolecular forces are the forces which cause attraction in the absence of chemical bonding. We are all aware of this attraction, since these forces appear in our everyday lives: intermolecular forces are, for example, responsible for the sticking together of snowballs and the appearance of surface tension. The formation of water droplets is a direct consequence of the existence of intermolecular forces between the water molecules.

These forces, which are about two orders of magnitude weaker than chemical bonds, are responsible for many phenomena in science, including some of the properties of real gases, as below:

- Van der Waals discovered in 1873 that real gases deviate from the ideal-gas equation. He concluded that the existence of attractive intermolecular forces results in a reduction of the measured pressure. This resulted in the proposal of an equation describing real gases very accurately, which is known today as the van der Waals equation:
  $(p + a/V^2) (V - b) = NkT$,
  with p, V, N, k, and T being the pressure, the volume, the number of molecules, the Boltzmann constant and the temperature of the gas, respectively; a and b are parameters which characterize the deviation from an ideal gas. The constant b has to be introduced since each molecule has a finite size, such that the volume of the gas cannot be compressed to zero. The constant a is a consequence of the attractive intermolecular forces that result in a reduction in the pressure.
- The transport properties of gases, such as viscosity, thermal conductivity and diffusion depend on the intermolecular forces.
- Intermolecular forces also effect the properties of condensed phases: A description of molecular solids and liquids requires knowledge of the intermolecular forces involved. In particular, the structure of molecular solids depends on the anisotropy of the potential. Our understanding of liquids and solvation has also been greatly increased with better characterization of the intermolecular potential. The special properties of water and ice

are explained by hydrogen bonding, a special case of a particularly strong
intermolecular attraction.

- In the case of molecular adsorption, the interaction of molecules with surfaces has to be described in terms of an intermolecular potential. A further
example is the description of zeolites, which can act as catalysts. In this
case the intermolecular forces between adsorbed molecules and the zeolite
lattice are important in determining the properties of the system.

- The biological activity of molecules is strongly influenced by intermolecular forces. In particular, the hydrogen bond, which has a strength that is
intermediate between the chemical and the weaker van der Waals bonds, is
highly directional and often determines the structure of biologically active
centers. However, these bonds can be opened with a modest amount of
energy, which allows changes in what would otherwise be a stable configuration. The duplication of the genetic code is accomplished in precisely this
manner. The well-known double helix of DNA (shown in Fig. 1.1) is held
together by hydrogen bonds. In the so-called Watson–Crick DNA structure
the guanine (G)–cytosine (C) and adenine (A)–thymine (T) pairs of bases
are bound by double or triple hydrogen bonds. For the duplication of the
code, the bonds open and each half is completed again with the corresponding bases. However, DNA is only one of many important and prominent
examples. In protein chains, for example, the secondary and tertiary structures of the peptides are stabilized by intermolecular interactions. The
dynamics of protein folding which is a hot topic in biology is governed
by intermolecular interactions. In chemistry the new field "supramolecular
chemistry" uses the concepts of intermolecular forces for the design of new
molecular complexes [170, 197].

From these few examples one can see the importance of an accurate description of intermolecular forces. At present, a full theoretical (ab initio)
description of the more complicated systems is still not possible; it is, therefore, highly desirable to obtain reliable models for the description of the
interactions. These models need to be accurately tested on relatively simple
prototype systems before they can be extended to more complex systems.

New experimental techniques have recently allowed detailed studies of
prototype systems, the goal of these studies being to gain new insight into
the associated intermolecular forces. In chemistry, we study molecules, which
are configurations of atoms bound by a certain amount of negative energy.
The molecule is stable if the binding energy exceeds the mean thermal energy,
which depends upon the temperature. As we go to lower and lower temperatures we find that any pair of molecules or atoms (with a few exceptions,
e.g. $^3\text{He}_2$ and NaHe) will eventually become stabilized solely because of the
weak intermolecular interactions.

Spectroscopic studies of these small complexes allow detailed study of
the intermolecular forces. It is the objective of this book to provide an introduction to the spectroscopy of small clusters that will provide the reader

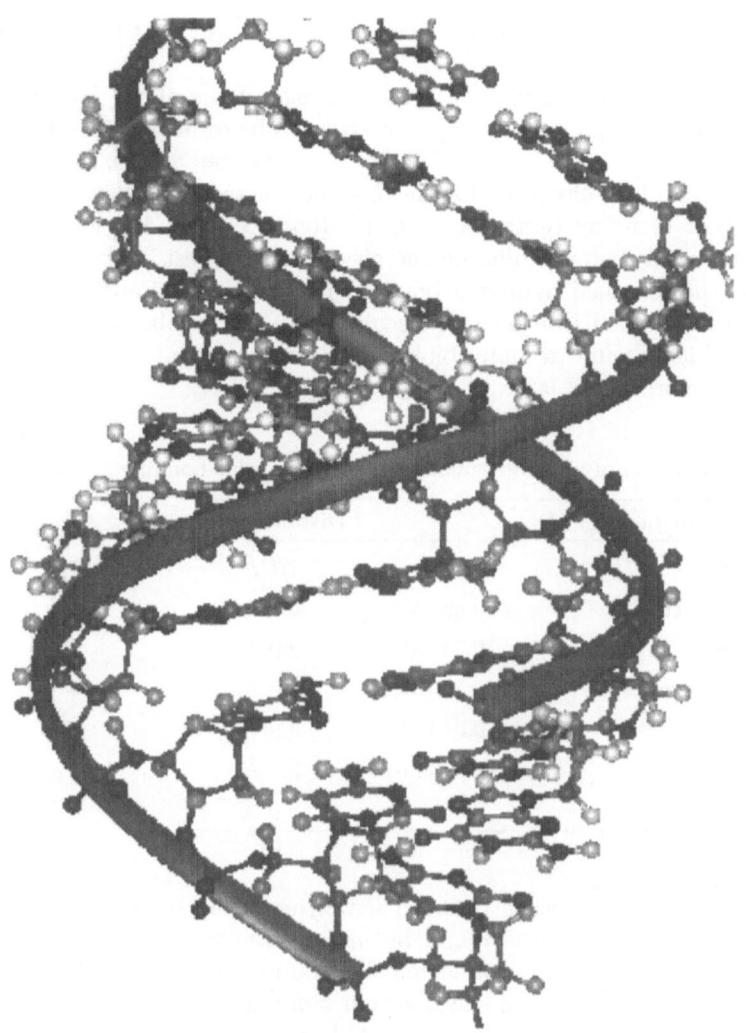

Fig. 1.1. Computer model of the DNA helix

with an appreciation of the tremendous progress that has been made in our understanding of small molecular complexes.

In chapter 2 and 3 the general concepts required to understand inter-molecular forces are introduced. The chapters 4–7 will give an overview of the experimental techniques that are currently being used to obtain detailed information about intermolecular potentials. In the third part of the book – chapter 8 and 9 – two prototype systems are discussed in detail.

1.2 Intramolecular – Versus Intermolecular Interactions

Van der Waals interactions comprise a broad range of the attractive forces that occur between any pair of molecules or atoms. The resulting weak bonding has to be distinguished from chemical bonding between atoms; the latter is described as covalent electron sharing or ionic electron transfer and has been the subject of many textbooks (e.g. [9, 105]). The binding energy is weak compared with chemical binding energies. The strongest intermolecular interaction is the so-called hydrogen bond. A molecule containing a hydrogen atom can be bonded to a polar molecule by intermolecular forces. This special case, which occurs in many biologically important interactions, will be discussed in more detail later.

Table 1.1. Comparison of intramolecular and intermolecular bonding

Molecule	Binding energy (D_e)		Distance (R_e)
CO	90500 cm^{-1}	11.2 eV	1.1 Å
Cs$_2$	3200 cm^{-1}	0.40 eV	4.5 Å
(H$_2$O)$_2$	1900 cm^{-1}	0.24 eV	2.0 Å
(NH$_3$)$_2$	1000 cm^{-1}	0.12 eV	3.4 Å
Ar–CO	110 cm^{-1}	0.014 eV	3.9 Å
Ne$_2$	30 cm^{-1}	0.0036 eV	3.1 Å
He$_2$	7.6 cm^{-1}	0.0009 eV	3.0 Å

Table 1.1 gives a comparison between some chemically bound molecules and some complexes which are bound by intermolecular interactions. The table contains the two extremes of chemically bound diatomics: CO, with a binding energy of 90 000 cm^{-1} and Cs$_2$, with a binding energy of 3200 cm^{-1}; the typical values are in between. We see that the strongest hydrogen-bonded complexes (e.g. (H$_2$O)$_2$) have binding energies of order 1900 cm^{-1}, which is only 1.7 times lower than that of the weakest "chemically" bound molecule (Cs$_2$). The weakest intermolecular bond has a binding energy of only 7.6 cm^{-1}, for the helium dimer. Although the line between chemical and van der Waals bonding is clearly blurred, bonds of the latter type are typically 100 times weaker than chemical interactions. This is also reflected in the corresponding bond lengths. Indeed, typical chemical bond lengths are in the range of 1.5 Å, whereas typical van der Waals bond lengths are in the range of 3–4 Å. Once again, the line between the two is not sharp. For example, the bond length for the chemical bond in Cs$_2$ is longer than the bond length for the weakly bound Ar–CO molecule. If we wish to distinguish between chemical and intermolecular bonds, we have to take the atomic radii into

account. Two atoms bonded by chemical interactions will have a distance between them which is less than the sum of the atomic radii. For Cs, with a large atomic radius of about 2.8 Å, the intramolecular distance is still smaller than the atomic diameter. For van der Waals interactions, it is generally true that the distance between two atoms is larger than the sum of the atomic radii. For example, the binding length of Ne_2 is approximately three times the corresponding atomic radius. Hydrogen-bonded complexes are intermediate between these two cases, with the bond length typically on the order of 2 Å, namely two times larger than that of the intramolecular bonds involved. A characteristic of intermolecular forces is that they are attractive over large distances, which explains the binding. However, for small distances, when the atomic radii overlap, the interaction becomes repulsive. We see that intermolecular interactions are weak compared with chemical interactions. As a consequence, in molecular complexes which are bonded by intermolecular interactions, the intramolecular distances and binding energies can be considered to be unchanged. This assumption has been confirmed experimentally.

2. Intermolecular Interactions

2.1 General Description

First, some general principles summarizing the general knowledge about intermolecular potentials will be presented. Chapter 3 gives a more detailed, quantitative description of intermolecular interactions. Textbooks discussing this subject include [125, 162, 182]. Later on I shall give an overview of the work of other groups, related to semiempirical potentials.

The Hamiltonian of a molecular complex is the sum of the kinetic and potential energy of all electrons and nuclei. In the so-called Born–Oppenheimer approximation the nuclear motion is separated from the electronic motion. This is a reasonable assumption if the lighter electrons follow instantaneously the motion of the nuclei. As a result any influence of the nuclear motion on the electronic energy is separated and treated as an intramolecular potential field $V(r)$ in which the nuclei move [85]. If we want to discuss intermolecular interactions we have to introduce the concept of "intermolecular potentials". The intermolecular potential is used to describe the binding between any pair of molecules (A, B) or atoms. For the simple case of an atom–atom pair, the binding energy is given by the difference between the energy of the complex and that of the two separated atoms:

$$V_{AB}(R) = E_{AB}(R) - E_A - E_B , \qquad (2.1)$$

where R is the distance between atom A and atom B. However, if one or both of the partners are molecules, the potential will depend on additional coordinates describing the relative orientation of the molecules with respect to one another. The simplest example of an atom–diatomic complex is shown in Fig. 2.1. R, r and θ denote the distance between the atom and the center of mass of the molecule, the intramolecular distance and the angle between the intermolecular and intramolecular axes, respectively. The potential field in which the nuclei move can now be described as $V(r, R, \theta)$. Similarly to the Born–Oppenheimer approximation, we can separate the intermolecular motion from the intramolecular motion by stating that the intramolecular distance is to a very good approximation not influenced by the intermolecular interaction. This implies that we separate the intramolecular motion from the much weaker intermolecular motion.

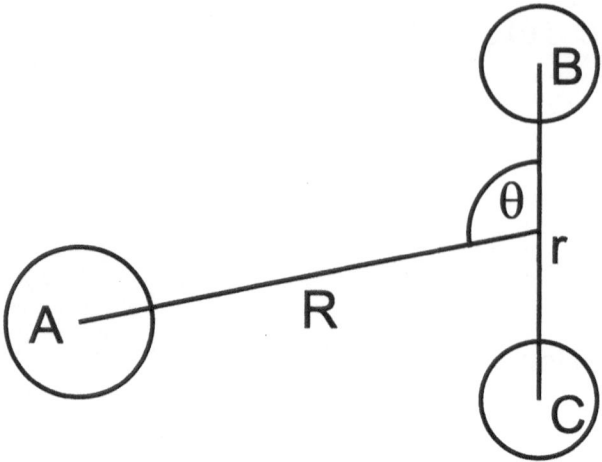

Fig. 2.1. Example of an atom–diatomic complex showing the coordinates R, r and θ of the intermolecular potential

Within this approximation the intramolecular distance can often be fixed and regarded as a constant. Within this approximation, therefore, the intermolecular potential depends on only two parameters: R and θ; the potential is actually represented by a two-dimensional surface. In more complex systems describing bimolecular complexes the number of parameters must be further increased to include angles describing the orientation of the two molecules relative to one another. Examples of an atom–diatomic complex and a bimolecular complex will be discussed in more detail later. Before doing that, however, it is helpful to consider the various contributions to the intermolecular interaction.

2.2 The Complete Hamiltonian

In this section the Hamiltonian for a complex consisting of the two monomers A and B is given. The Hamiltonian consists of parts describing the intramolecular and intermolecular interactions. The two components will then be separated. Since the intermolecular interaction is weak, we can use perturbation theory for a description. More specifically, the Hamiltonian of the complex AB can be described in the following way:

$$H = H_A + H_B + V_{AB} \,, \tag{2.2}$$

where H_A describes the free monomer A, and H_B describes the free monomer B. V_{AB} contains all parts describing the intermolecular interaction between monomers A and B. Using perturbation theory, we begin with an eigenvector of the Hamiltonian H_0, which describes the system in the absence of the perturbation V_{AB}:

$$H_0 = H_A + H_B \;.$$

As the eigenvector ψ_0 we choose the product of the unperturbed monomer wave functions, $\psi_0 = \psi_A \psi_B$, which fulfills the following condition:

$$H_0 \psi_0 = (H_A + H_B)\,\psi_0 = (E_A + E_B)\,\psi_0 \;,$$

and therefore $E_0 = E_A + E_B$.

However, taking this as the eigenfunction of the entire system has one inherent problem. The eigenfunction is no longer completely antisymmetrized regarding the exchange of electrons, as is required for fermions. This will become clear if we write the wave function in more detail:

$$\psi_0 \left(r_1, \ldots, r_{n_A}, r_{n_A+1}, \ldots, r_{n_A+n_B}\right)$$
$$= \psi_A \left(r_1, \ldots, r_{n_A}\right) \psi_B \left(r_1, \ldots, r_{n_B}\right). \tag{2.3}$$

This implies that

$$\psi \left(r_1, \ldots, r_{n_A}, r_{n_A+1}, \ldots, r_{n_A+n_B}\right)$$
$$= -\psi \left(r_2, r_1, r_3, \ldots, r_{n_A}, r_{n_A+1}, \ldots, r_{n_A+n_B}\right)$$
$$= -\psi \left(r_1, \ldots, r_{n_A}, r_{n_A+2}, r_{n_A+1}, r_{n_A+3}, \ldots, r_{n_A+n_B}\right),$$

since each of the monomer wave functions is completely antisymmetrized. However, if we exchange electrons between monomers A and B, we find that

$$\psi \left(r_1, \ldots, r_{n_A}, r_{n_A+1}, \ldots, r_{n_A+n_B}\right)$$
$$\neq -\psi \left(r_{n_A+1}, r_2, \ldots, r_{n_A}, r_1, r_{n_A+2}, \ldots, r_{n_A+n_B}\right).$$

As long as we consider regions where R is large compared with the intramolecular distances, we can still use normal perturbation theory in combination with ψ_0. For typical R values, overlap terms and therefore the exchange of electrons between the different monomers can be neglected. We shall therefore continue using ψ_0 as the appropriate zero-order wave function and work out the different interaction terms in first- and second-order perturbation theory. Later on, however, when we discuss the interaction for small R values, we shall come back to this problem. The full Hamiltonian is given by:

$$H = H_0 - \sum_{i=1}^{n_A} \sum_{\beta=1}^{N_B} \frac{Z_\beta}{r_{i\beta}} - \sum_{j=1}^{n_B} \sum_{\alpha=1}^{N_A} \frac{Z_\alpha}{r_{j\alpha}} + \sum_{i=1}^{n_A} \sum_{j=1}^{n_B} \frac{1}{r_{ij}} + \sum_{\alpha=1}^{N_A} \sum_{\beta=1}^{N_B} \frac{Z_\alpha Z_\beta}{r_{\alpha\beta}}$$
$$= H_0 + V_{AB} \;, \tag{2.4}$$

with n_X, N_X, Z_X and r_{xy} being the number of electrons of monomer X, the number of atoms of monomer X, the charge of each nucleus of monomer X and the distance between x and y, respectively.

2.3 Electrostatic Energy

We shall now consider the corrections from first-order perturbation theory, for which the energy correction is given by $E_1 = \langle \psi_0^0 | V_{AB} | \psi_0^0 \rangle$, with ψ_0 describing the eigenfunction corresponding to the lowest eigenvalue E_0^0 of H_0. Therefore, $\psi_0^0 = \psi_A^0 \psi_B^0$, with $H_X \psi_X^0 = E_X^0 \psi_X^0$, in which E_X^0 denotes the ground state of monomer X. If we work out the integrals we obtain the following result:

$$
\begin{aligned}
E_1 = \iiiint & \psi_A^0(x)^* \psi_B^0(y)^* \frac{\hat{\rho}_A(r,x)\hat{\rho}_B(r',y)}{|r - r'|} \psi_A^0(x)\psi_B^0(y) \\
& \times \, d\tau_x \, d\tau_y \, d^3 r \, d^3 r' \, ,
\end{aligned}
\tag{2.5}
$$

where $\hat{\rho}_X$ is the charge density operator of monomer X, given by

$$
\hat{\rho}_X(r,x) = \left(\sum_{\alpha=1}^{N_X} Z_\alpha \, \delta(r - x_\alpha) - \sum_{i=1}^{n_X} \delta(r - x_i) \right).
$$

The charge density of the unperturbed monomer X is given by

$$
\rho_X(r) = \int \psi_X^0(x)^* \rho_X(r,x) \psi_X^0(x) \, d\tau_x \, .
\tag{2.6}
$$

If we insert this into (2.5) we find

$$
E_1 = \int \rho_A(r) \int \frac{\rho_B(r')}{|r - r'|} d^3 r \, d^3 r'.
\tag{2.7}
$$

We can see from the above equation that this energy is the result of the electrostatic interaction of the charges of the unperturbed monomer A with the charges of the unperturbed monomer B. Correspondingly, the energy E_1 will be called E_{pol} in the following. E_{pol} can be written as the product of the charge density of monomer A with the electrostatic potential of the charge distribution of monomer B ($V_B(r)$):

$$
E_{\text{pol}} = \int \rho_A(r) V_B(r) \, d^3 r \, ,
\tag{2.8}
$$

where

$$
V_B(r) = \int \frac{\rho_B(r')}{|r - r'|} d^3 r' \, .
\tag{2.9}
$$

We can now make use of the fact that we are only interested in solutions for which the distance between monomer A and monomer B is much larger than the intramolecular distances, namely, the asymptotic behavior of the solution. For this purpose we rewrite $|r - r'|$ in the following way: $|r - r'| = |r_A - r_B - R|$, as illustrated in Fig. 2.2. $|R|$ is the distance between the centers of mass of the two monomers; r_A and r_B are the coordinates of the nuclei and electrons

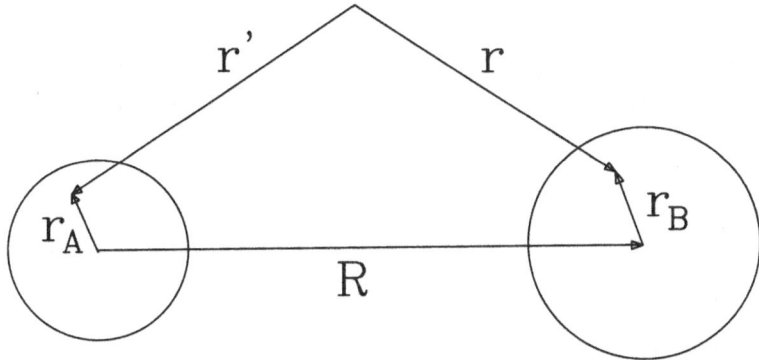

Fig. 2.2. New coordinate system for the multipole expansion

in new coordinate systems with the centers of mass of monomers A and B as their origins. In regions with $|R| > |(r_A + r_B)|$, which means in regions where charge overlap is negligible, we can use the following relation (see also [105]):

$$\frac{1}{|r_A - r_B - R|}$$
$$= \sum_{l_A=0}^{\infty} \sum_{l_B=0}^{\infty} \sum_{m=-l_<}^{l_<} d_{l_A l_B}^m \frac{r_A^{l_A} r_B^{l_B}}{R^{l_A+l_B+1}} Y_{l_A}^m(\theta_a, \phi_a) Y_{l_B}^{-m}(\theta_b, \phi_b) , \qquad (2.10)$$

where $l_<$ is the smaller of l_a and l_b, θ_X and ϕ_X are the angles which describe the orientation of the vector r_X in the coordinate system of monomer X, and $d_{l_A l_B}^m$ is a constant, defined as

$$d_{l_A l_B}^m = \frac{4\pi}{\sqrt{(2l_A+1)(2l_B+1)}} \frac{(l_A+l_B)!(-1)^{l_A+m}}{\sqrt{(l_A+m)!(l_A-m)!(l_B+m)!(l_B-m)!}} .$$

The bipolar expansion (2.10) will be used to transform (2.5) into a more convenient form, in which the energy resulting from the electrostatic interaction can be related to physical properties of the monomer. We can now separate the two integrals in (2.5), leading to the following equation:

$$E_{\text{pol}}$$
$$= \sum_{l_A=0}^{\infty} \sum_{l_B=0}^{\infty} \sum_{m=-l_<}^{l_<} d_{l_A l_B}^m \frac{\sqrt{(2l_A+1)(2l_B+1)}}{4\pi R^{l_A+l_B+1}} \int f(r_A)\, d\tau_A \int f(r_B)\, d\tau_B , $$
$$\qquad (2.11)$$

with $f(r_X) = r_X^{l_X} \rho_X(r_X) Y_{l_X}^m(\theta_X, \phi_X) \sqrt{4\pi/(2l_x+1)}$.

We now introduce the electric multipole moments of the charge distribution, which are defined as

$$Q_X^{lm} = \sqrt{\frac{4\pi}{2l+1}} \int r_X^{l_X} \rho_X(r_X) Y_l^m(\theta_A, \phi) \, d\tau_X \; . \tag{2.12}$$

If we take a closer look at the lowest-order multipole moments, we obtain

$$Q_X^{00} = \int \rho_X(r_X) \, d\tau_X \; ,$$

which corresponds to the net charge of monomer X and is zero for a neutral monomer.

$$Q_X^{10} = \int z_X \rho_X(r_X) \, d\tau_X$$

corresponds to the z component of the dipole moment (μ_z), and

$$Q_X^{1\pm1} = \frac{1}{\sqrt{2}}(\mu_x \pm i\mu_y) \; .$$

If we introduce the definition (2.12) into (2.11), we obtain the following expression:

$$E_{\text{pol}} = \sum_{l_A=0}^{\infty} \sum_{l_B=0}^{\infty} \sum_{m=-l_<}^{l_<} \frac{(-1)^{l_A+m}(l_A + l_B)!}{\sqrt{(l_A + m)!(l_B + m)!(l_A - m)!(l_B - m)!}}$$
$$\times \frac{Q_A^{l_A m} Q_B^{l_B - m}}{R^{l_A+l_B+1}} \; . \tag{2.13}$$

This is the usual form for the description of the electrostatic contribution to the intermolecular interaction. We can easily deduce from this expression the asymptotic behavior (as $R \to \infty$). For a neutral monomer, the first non-vanishing term describes the dipole–dipole interaction, which is proportional to R^{-3}. The quadrupole–dipole interaction varies as R^{-4}, the quadrupole–quadrupole as R^{-5} and so on.

For polar molecules, the dipole–dipole interaction is a significant contribution to the intermolecular interaction. If we consider, for example, the case of $(\text{HF})_2$ (see Fig. 2.3), the dipole–dipole interaction can be described simply by

$$E_{\text{pol}} = -\frac{2\mu_{\text{HF}} \cos \varphi \, \mu_{\text{HF}}}{R^3} \; ,$$

where φ is the angle between the two HF monomers and μ_{HF} describes the electric dipole moment of the HF monomer.

As a result of this interaction the two monomers will tend to align with the two dipoles pointing in the same direction, since this is the energetically most favorable position. However, the experimentally determined structure of $(\text{HF})_2$ is quite different from this, with φ being 108° [46]. The deviation from a linear structure can be explained if we consider the quadrupole–quadrupole

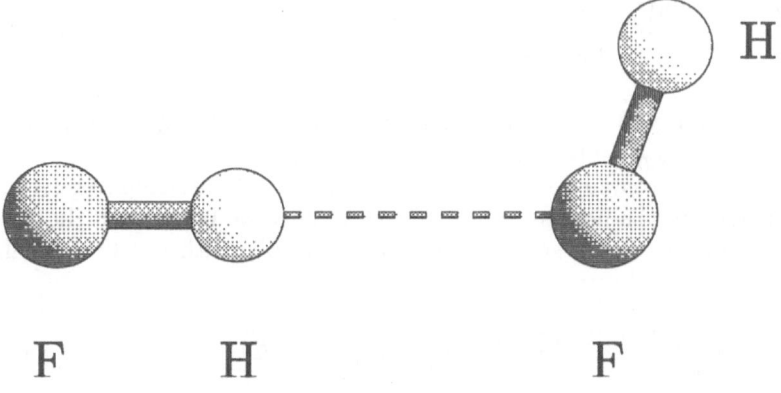

H

F   H   F

Fig. 2.3. Structure of $(HF)_2$

interaction, which favors a T-shaped structure. The final structure ($\varphi = 108°$) is primarily a compromise between the maxima in the dipole–dipole and the quadrupole–quadrupole interaction energies. Indeed, the experimental geometry agrees very well with the value calculated taking into account only the total electrostatic energy of the dimer ($\varphi = 112°$). This example shows that the dipole–dipole term and, more generally, the electrostatic part of the intermolecular potential can be attractive (for $\varphi = 0°$) or repulsive (for $\varphi = 180°$), depending sensitively on the orientation of the two monomers. We shall see in the following section that the second-order terms can also contribute substantially to the overall binding energy, although they are not as directional. Therefore, in cases where the electrostatic imbalance in the molecule is above a certain level, this will determine the structure of the dimer, which will be such that the electrostatic interaction will be attractive and as large as possible. This is normally expected to be the case for hydrogen-bonded structures; however, it fails for $(NH_3)_2$, as we shall see later on. A more detailed discussion of these issues can also be in [19, 21].

2.4 Induction Energy

We shall now consider the contributions resulting from second-order perturbation theory. In the previous section we introduced ψ_0^0, which is an eigenfunction of the Hamiltonian H_0 with the eigenvalue $E_0 = (E_A + E_B)_0$, which implies that ψ_0^0 is the eigenfunction describing the ground state of H_0. In this section we introduce the eigenfunction ψ_0^k of H_0, which corresponds to the eigenvalue E_0^k, with $k \neq 0$. We can write $E_0^k = E_A^m + E_B^n$, with either n or m or both $\neq 0$. The correction resulting from second-order perturbation theory can then be written as

$$E_2 = -\sum_{k \neq 0} \frac{\langle \psi_0^0 | V_{AB} | \psi_0^k \rangle \langle \psi_0^k | V_{AB} | \psi_0^0 \rangle}{E_0^k - E_0^0} .$$
(2.14)

In general the sum will include all eigenfunctions $\psi_0^k = \psi_A^n \psi_B^m$ with the exception of $\psi_0^0 = \psi_A^0 \psi_B^0$. In this section we shall investigate all terms for which n or m is equal to zero and deduce a general expression. The remaining terms will be discussed in the next section. For convenience we begin with the calculation of the terms in the sum for which $\psi_0^k = \psi_A^0 \psi_B^n$, with $n \neq 0$, and call this part E_{ind}:

$$E_{ind} = -\sum_{n \neq 0} \frac{\langle \psi_A^0 \psi_B^0 | V_{AB} | \psi_A^0 \psi_B^n \rangle \langle \psi_A^0 \psi_B^n | V_{AB} | \psi_A^0 \psi_B^0 \rangle}{E_B^n - E_B^0} .$$
(2.15)

In more detail, with $E_B^n - E_B^0 = \epsilon_B^n$ and using (2.9),

$$E_{ind} = -\sum_{n \neq 0} \frac{1}{\epsilon_B^n} \int V_A(r) \psi_B^0(y)^* \hat{\rho}(r,y) \psi_B^n(y) \, d\tau_y \, d^3r$$

$$\times \int V_A(r') \psi_B^n(y')^* \hat{\rho}(r',y') \psi_B^0(y') \, d\tau_{y'} \, d^3r' .$$
(2.16)

This can be written as

$$E_{ind} = -\frac{1}{2} \iint V_A(r) \alpha_{r,r'}^B V_A(r') \, d^3r \, d^3r' ,$$
(2.17)

if we define the polarizibility $\alpha_{r,r'}^B$ as

$$\alpha_{r,r'}^B = 2 \sum_{n \neq 0} \iint \frac{\psi_B^0(y)^* \hat{\rho}(r,y) \psi_B^n(y) \psi_B^n(y')^* \hat{\rho}(r',y') \psi_B^0(y')}{\epsilon_B^n} \, d\tau_y \, d\tau_{y'} .$$

Equation (2.17) describes the induction energy. This is the energy of the molecule B under the influence of an external field, which is described by the potential $V_A(x,y)$. The electric potential is created by the permanent charge distribution of monomer A.

We now want to study the asymptotic behavior of this contribution. To do this, we shall have to use the result of the expansion of $1/(|r - r'|) = 1/(|r_A - r_B - R|)$. Inserting (2.10), we can separate the expression above into a product of terms depending solely on r_A, r_B and R. We further use the fact that

$$Q_A^{lm} \propto \int \rho_A(r_A) r_A^{l_A} Y_{l_A}^m(\theta_A, \phi_A) \, d\tau_{r_A}$$

and

$$Q_A^{l'm'} \propto \int \rho_A(r_{A'}) r_{A'}^{l_{A'}} Y_{l_{A'}}^{m'}(\theta', \phi') \, d\tau_{r_B} .$$

E_{ind} is then proportional to the following expression:

$$
\sum_{l_A, l_B = 0}^{\infty} \sum_{m=-l_<}^{m=l_<} \sum_{l_{A'}, l_{B'} = 0}^{\infty} \sum_{m'=-l'_<}^{m'=l'_<} \frac{Q_A^{lm} Q_A^{l'm'}}{R^{l_A+l_B+1+l_{A'}+l_{B'}+1}}
$$

$$
\times \iint r_B^{l_B} r_{B'}^{l_{B'}} \alpha_{r_B, r_{B'}}^{B} \, d\tau_{r_B} \, d\tau_{r_{B'}} \, . \tag{2.18}
$$

This equation describes the asymptotic behavior for large R values. However, we have to take a closer look at the integral in (2.18). Using the definition of α, we can separate the two integrals and write them as the product of two terms. One of them can be written as

$$
\langle \psi_B^0(y)^* | \int \hat{\rho}(r_B, y) r_B^{l_B} \, d^3 r_B \, | \psi_B^n(y) \rangle \, .
$$

The eigenfunctions ψ_B^l constitute an orthonormal basis. For $l_B = 0$ the integral $\int \hat{\rho}(r_B, y) \, d^3 r_B$ is a constant and the whole expression is zero, since $n \neq 0$. We therefore only have nonzero contributions in (2.18) if $l_B, l_{B'} \neq 0$. The parameters l_A and $l_{A'}$ will depend on the multipole moments of monomer A.

More generally, we can now add up all contributions to the correction to E_0 that include $\psi_0^k = \psi_A^m \psi_B^n$ with either m or n equal to zero. We have now calculated the terms with $m = 0, n \neq 0$; in the same way we can obtain all terms with $m \neq 0, n = 0$, which also contribute to the induction energy. We can summarize both in the following equation:

$$
E_{ind} = - \sum_{n, n'=2}^{\infty} C_{ind}^{nn'} R^{-n-n'} \, , \tag{2.19}
$$

where $n = l_A+l_B+1$, $n' = l_{A'}+l_{B'}+1$ and $C_{ind}^{n,n'}$ is a constant, which depends on the multipole moment of monomer 1 and the polarizility of monomer 2. From the discussion above, it is obvious that we have the restriction that if $l_A, l_{A'} = 0$ then $l_B, l_{B'} \geq 1$, and if $l_B, l_{B'} = 0$ then $l_A, l_{A'} \geq 1$, which implies $n, n' \geq 2$. If one of the monomers possesses a free charge, which means that one of the monomers is an ion, the lowest order of the induction energy will vary as

$$
E_{ind} \propto \frac{1}{R^4} \, .
$$

This is in agreement with classical theory, which requires that

$$
E_{ind} = -E\mu_{ind} \propto \frac{1}{R^4} \, ,
$$

with μ_{ind} being the dipole moment induced by the electric field E generated by the monomer B. For neutral molecules the lowest order of multipole moment is Q^{10}. If one monomer possesses an electric dipole moment the induction energy will vary as $1/R^6$. In contrast to the electrostatic contribution, described in the previous section, the induction energy will contribute to the overall energy even if all multipole moments of monomer B are equal to zero. We can illustrate this for the complex Ar–HF, which is shown in Fig. 2.4.

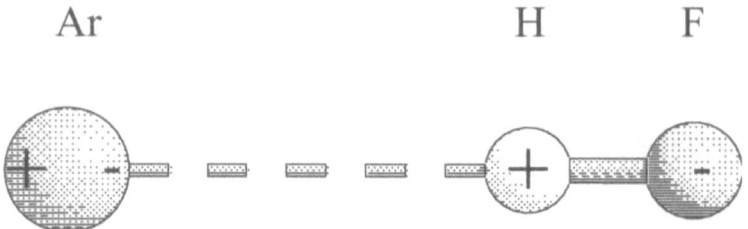

Fig. 2.4. The induced dipole–dipole interaction in Ar–HF

The electric dipole in HF induces a dipole in the argon atom, giving rise to a dipole-induced dipole interaction. The induced dipole will always point in the same direction as the permanent dipole moment, the interaction energy thereby reaching the maximum possible value. The induction energy is always attractive, since E_{ind} is always negative, as can be seen from (2.15). The induction energy is anisotropic, since it depends on the direction of the multipole moment. However, it is usually smaller than the electrostatic energy, except of course in a complex where one of the monomers has vanishing multipole moments.

2.5 Dispersion Energy

In this section we investigate the contributions resulting from second-order perturbation theory that were left out in the previous section: the parts with $\psi_0^k = \psi_A^m \psi_B^n$, with $m \neq 0$ and $n \neq 0$. This contribution is a pure quantum mechanical quantity for which no classical analog can be found. It is called the dispersion energy E_{disp}, which can be written as (see also (2.15))

$$E_{\text{disp}} = -\sum_{m \neq 0} \sum_{n \neq 0} \frac{\langle \psi_A^0 \psi_B^0 | V_{AB} | \psi_A^m \psi_B^n \rangle \langle \psi_A^m \psi_B^n | V_{AB} | \psi_A^0 \psi_B^0 \rangle}{\epsilon_A^m + \epsilon_B^n} . \quad (2.20)$$

Analogously to the derivation in the previous section, we can now make use of the multipole expansion (2.10) to separate the integrals. Further, we can insert the definition of the polarizibility α to determine the asymptotic behavior for $R \to \infty$. If we do so, we can deduce that the dispersion energy has the following proportionality:

$$E_{\text{disp}} \propto - \sum_{m,n\neq 0} \frac{\epsilon_A^m \epsilon_B^n}{\epsilon_A^m + \epsilon_B^n} \sum_{l_A,l_B=1}^{\infty} \sum_{m=-l_<}^{m=l_<} \sum_{l_{A'},l_{B'}=1}^{\infty} \sum_{m'=-l'_<}^{m'=l'_<}$$

$$\times \iiiint \frac{r_A^{l_A} r_B^{l_B} r_{A'}^{l_{A'}} r_{B'}^{l_{B'}} \alpha_{r_A,r_{A'}}^A \alpha_{r_B,r_{B'}}^B}{R^{l_A+l_B+1+l_{A'}+l_{B'}+1}} \, \mathrm{d}\tau_{r_B} \, \mathrm{d}\tau_{r_{A'}} \, \mathrm{d}\tau_{r_A} \, \mathrm{d}\tau_{r_{B'}} \ . \quad (2.21)$$

The sum starts at $l_A, l_B = 1$ instead of $l_A, l_B = 0$ for the same reason as discussed in the previous section. Any integral of the form $\langle \psi_X^0 | r_X^{l_X} \rho_X | \psi_X^m \rangle$ which is implicitly given in the expression above is zero for $l_X = 0$. We now have a similar functional form for the induction and dispersion energies, the only difference being the smallest possible values of n and n':

$$E_{\text{disp}} = - \sum_{n,n'=3}^{\infty} C_{\text{disp}}^{nn'} R^{-n-n'} \ , \quad (2.22)$$

with $n = l_A + l_B + 1$ and $n' = l_{A'} + l_{B'} + 1$. $C_{\text{disp}}^{n,n'}$ depends solely on the polarizibilities of monomers 1 and 2, so that it follows an R^{-6} dependence for any molecule.

Ar   Ar

Fig. 2.5. The induced-dipole–induced-dipole interaction in $(\text{Ar})_2$

We have an attractive interaction, which is also present between two monomers with no multipole moments ($Q^{lm} = 0$), e.g. between two atoms. Although there is no classical equivalent to this attractive interaction, we can visualize it using a picture similar to that give above for the dipolar interactions. To do so we may consider the case of $(\text{Ar})_2$, which is displayed in Fig. 2.5. As is also indicated in (2.21), the dispersion interaction arises because multipole moments are induced simultaneously in both atoms and can interact with one another. The appearance of instantaneous multipole moments can be understood in terms of quantum mechanical charge fluctuations. The induced multipole moments will be correlated, thereby making the interaction energy as large as possible. A possible configuration, in which the two instantaneous dipole moments are lined up, is shown in Fig. 2.5. If the dipole moment of monomer A formed by fluctuating electrons is aligned with the vector R, the dipole moment of monomer B will point in the same direction. The dispersion energy is more isotropic than the electrostatic and induction energies, since the fluctuating multipole moments can appear in any direction and are not as directional as the permanent multipole moments. For monomers where all multipole moments are zero (e.g. noble gases), this interaction is the only contribution to the attractive van der Waals interaction. Even for polar molecules, however, the dispersion energy can be a significant contribution to the overall binding energy.

2.6 Overlap–Exchange Repulsion

So far, we have discussed only the contributions to long-range interactions. As was pointed out earlier, the assumption was made that $\psi_0 = \psi_A \psi_B$, which is not valid for $R \rightarrow 0$. We shall now consider the contributions to the short-range interactions. This contribution to the potential is really a combination of two parts: exchange, which means that electron motions can extend over both molecules, and repulsion, which arises from the fact that, in terms of molecular orbitals, an antibonding orbital will be populated.

When combined, these give rise to a repulsive contribution, since at short range molecules will repel one another. If we exclude any chemical binding, when the charge distributions of the monomers begin to overlap there will be a strong repulsive contribution. More precisely, the repulsive contribution can be calculated if we use the correct description for $R \rightarrow 0$. To do this, we must overcome the problem that ψ_0 is not fully antisymmetrized. As a first step, we construct the fully antisymmetrized wave function:

$$\frac{1}{N!} \sum_{\alpha=1}^{N!} \epsilon_\alpha P_\alpha \psi_0 \,,$$

with $N!$ being the number of all possible permutations of electrons (P_α), and ϵ being $+1$ for even permutations and -1 for odd permutations. However, this wave function is not an eigenfunction of $H_0 = H_A + H_B$! For small R values, we therefore cannot apply conventional perturbation theory, as was done in the previous sections. Instead, for a correct treatment a new type of perturbation theory, the Murrell–Shaw–Musher–Amos perturbation theory, has to be used. (This is discussed in detail in [137, 138].) The result of this treatment is that the correct result of first-order perturbation theory can be written as

$$E_1 = \frac{1/N! \sum_{\alpha=1}^{N!} \epsilon_\alpha \langle P_\alpha \psi_0 | V_{AB} | \psi_0 \rangle}{1/N! \sum_{\alpha=1}^{N!} \epsilon_\alpha \langle P_\alpha \psi_0 | \psi_0 \rangle} \,. \tag{2.23}$$

This equation consists of two parts. The first can be attributed to the electrostatic energy, which was discussed earlier in detail. The second consists of correction terms, which appear only in the correct solution. We can therefore write

$$E_1 = E_{\text{pol}} + E_{\text{corr}} \cdot$$

The correction term turns out to be the overlap–exchange energy, which leads to a strongly repulsive interaction for small R values. We have to separate the permutations P_α in (2.23) into those which are within monomer A or B and those which exchange electrons between the monomers, here denoted as P^{AB}:

$$\sum_{\alpha=1}^{N!} \epsilon_\alpha P_\alpha = \sum_{\alpha_A=1}^{N_A!} \sum_{\alpha_B=1}^{N_B!} \epsilon_{\alpha_A} \epsilon_{\alpha_B} P^A_{\alpha_A} P^B_{\alpha_B} + \sum_{(\alpha=N_A!+N_B!)}^{N!} \epsilon_\alpha P^{AB}_\alpha \cdot$$

Only the final result is given here. Using

$$\tilde{P} = \sum_{(\alpha=N_A!+N_B!)}^{N!} \frac{\epsilon_\alpha P^{AB}_\alpha}{N_A! N_B!} \, ,$$

we obtain

$$
\begin{aligned}
E_1 &= \frac{\langle \psi_0 + \tilde{P}\psi_0 | V_{AB} | \psi_0 \rangle}{\langle \psi_0 + \tilde{P}\psi_0 | \psi_0 \rangle} \\
&= \frac{\langle \psi_0 | V_{AB} | \psi_0 \rangle}{1 + \langle \tilde{P}\psi_0 | \psi_0 \rangle} + \frac{\langle \tilde{P}\psi_0 | V_{AB} | \psi_0 \rangle}{1 + \langle \tilde{P}\psi_0 | \psi_0 \rangle} \\
&= \langle \psi_0 | V_{AB} | \psi_0 \rangle + \frac{\langle \tilde{P}\psi_0 | (V_{AB} - \langle \psi_0 | V_{AB} | \psi_0 \rangle) | \psi_0 \rangle}{1 + \langle \tilde{P}\psi_0 | \psi_0 \rangle} \\
&= E_{\text{pol}} + E_{\text{exov}} \cdot
\end{aligned}
$$

(2.24)

For the correction term (E_{exov}), we can write

$$
\begin{aligned}
E_{exov} = {} &\langle \tilde{P}\psi_0 | V_{AB} | \psi_0 \rangle - E_{pol} \langle \tilde{P}\psi_0 | \psi_0 \rangle \\
&- \langle \tilde{P}\psi_0 | V_{AB} | \psi_0 \rangle \langle \tilde{P}\psi_0 | \psi_0 \rangle + O(\langle \tilde{P}\psi_0 | \psi_0 \rangle) \cdot
\end{aligned}
$$

(2.25)

This term is called the exchange–overlap energy since it contains terms which describe the exchange energy, $\langle \tilde{P}\psi_0 | V_{AB} | \psi_0 \rangle$, and parts which are proportional to the overlap of the wave functions, $\langle \tilde{P}\psi_0 | \psi_0 \rangle$.

The overlap–exchange energy is a repulsive contribution to the intermolecular potential. For small R values we obtain a partial overlap of the wave functions, and since we are dealing here with two monomers which are not chemically bound (e.g. Ar–Ar), we obtain a repulsive force. There are of course also corrections to the terms derived by second-order perturbation theory, which are also repulsive; however, a correct treatment of these terms is beyond the scope of this book. It has been shown in many studies that the repulsive part of the potential can be described very accurately by an exponentially decaying function, the so-called Born–Mayer potential:

$$E_{\mathrm{exov}} = A \exp(-bR) \, , \tag{2.26}$$

with A and b being two adjustable parameters which depend on the relative angular orientation of the monomers [188]. This functional form has been used in all recent semiempirical studies of intermolecular interactions (e.g. [36, 90, 93, 146, 158, 169]). This way of describing the repulsive part of the potential has turned out to be very successful, even though it is not yet possible to find a strict derivation of A and b starting from an ab initio calculation.

At smaller distances the multipole expansion for the contributions previously discussed fails even when it converges. This is caused by the penetration of charges. The penetration energy describes the correction at short distances that arises from the finite extent of the charge distribution. It corrects the multipole expansion in regions in which $|R| > |(r_A + r_B)|$ does not hold. The correction cancels out all singularities that would arise from the multicenter expansion. This correction has to be taken into account in the electrostatic energy, the induction energy and the dispersion energy. In general, it is described by a so-called damping function [183]. The exact mathematical form will be described in more detail in the next section. The penetration correction does not influence the long-range behavior, but corrects the short-range forms of the different contributions.

2.7 Nonspherical Potentials

Since spherical interactions are already by far the best understood, the focus of the present work will therefore be on the study of nonspherical potentials.

A problem arises, especially for larger molecules, since the multipole expansion is convergent only for $|R| > |(r_A + r_B)|$, which means for R values that are large compared with the size of the molecule. Even for moderately sized molecules this means that the R value for which a multipole expansion is adequate depends on the orientation, since the molecule is large in one direction and small in the other. If we have an anisotropic charge distribution, a centrosymmetric approach is not justified. Instead, each contribution will be angulary dependent, which implies that all of the C_n parameters, as well as A and b, will show an angular dependence.

One approach for dealing with this is to use distributed multipoles, which means that several centers in the molecule are chosen, each describing a centrosymmetric charge distribution. These are chosen such that they are in physically meaningful locations and the overall sum gives the optimum description of the overall molecule. This turns out to be advantageous, especially in the case of large molecules.

The mathematical procedure is as follows. So far, we have considered single-center expansions, which means that for a description of the electrostatic, induction and dispersion energies we have used an expansion of

$1/(|r_x - r_y|)$ in terms of $1/R$. In this way we could describe the charge distribution of the monomer by a series of multipole moments, expanded at a single center in each monomer, corresponding to the center of mass. In the same way, the polarizabilities, which describe the induced charge densities, have been expanded and used for the calculation of the C_n^l parameters. The interaction was then described in terms of the interaction of multipoles with multipoles (2.13), polarizabilities with multipole moments (2.18) or polarizabilities with polarizabilities (2.21). For a rather extended, anisotropic charge distribution, accurate descriptions of the interaction energy require the inclusion of high-order terms for the accurate angular description of these interactions. However, as pointed out previously, a small number of centers is desirable.

If we consider an anisotropic charge distribution, e.g. for CO, the multipole moments can now be chosen such that the field generated by this charge distribution is equivalent to the field generated by the various multipole moments positioned at the center of mass of the CO molecule. It is obvious that, for a precise description of the angular dependence at close distances, high-order terms in the expansion of the anisotropy (described by $P_l(\cos\theta)$) are required. Instead, it is possible to describe the charge distribution as the sum of two nearly centrosymmetric charge distributions. In the same way, the dispersion interaction can be distributed over two bonds, which can also lead to the reduction of high-order terms in the description of the anisotropy of the dispersion interaction. This approach was first used in [44]. For an accurate description of the Ar–C_2H_2 potential, the distribution of the dispersion interaction over the length of the acetylene subunit was found to be essential [8]. Similar considerations, especially for the purpose of modeling intermolecular interactions between large molecules, have led to the development of theoretical concepts of distributed multipoles and distributed polarizabilities (e.g. [3, 66, 180, 198]).

After this description of the main contributions to the intermolecular potential, we shall use the results of this section to obtain a parameterization of the intermolecular potential. This will be then used for a semiempirical description of the potential-energy surface.

3. Semiempirical Potentials

3.1 Form of the Potential

In Chap. 2 we discussed the various attractive and repulsive contributions to the overall intermolecular interaction. The long-range asymptotic behavior of the attractive interactions can be represented as R^{-n} ($n \geq 3$), which decreases rather slowly compared with the repulsive contribution, which decreases rapidly, as $\exp(-bR)$. For large R values, the potential will therefore follow an R^{-n} dependence; for example, the interaction between two nonpolar molecules follows the well-known R^{-6} dependence. For small R values ($R \to 0$) there will be a strong repulsion. We now consider a functional form for the intermolecular potential, which is relatively simple and contains only a few, physically meaningful, adjustable parameters. In addition, we require that it gives an appropriate description for the two limiting cases $R \to \infty$ and $R \to 0$. The asymptotic behavior for $R \to \infty$ was discussed for the various different contributions in the previous chapter. If we summarize the various contributions obtained from second-order perturbation theory, we obtain the following equation:

$$
\begin{aligned}
V &= V_{\text{exov}} + V_{\text{pol}} + V_{\text{ind}} + V_{\text{disp}} \\
&= A \exp(-bR) - \sum_{n=3}^{\infty} \frac{C_{\text{pol},n}}{R^n} - \sum_{n=6}^{\infty} \frac{C_{\text{ind},n}}{R^n} - \sum_{n=6}^{\infty} \frac{C_{\text{disp},n}}{R^n} ,
\end{aligned}
\tag{3.1}
$$

where $C_{\text{pol},n}$ in the electrostatic term depends on the multipole moments of the monomer. For a noble-gas–molecule interaction this term will be zero. Equation (3.1) represents a convenient form for the intermolecular potential, with several angularly dependent parameters (A, b and C^n), which can be associated with physical quantities such as the multipole moment Q^{lm} and the polarizability of a monomer $\alpha_{r,r'}$. However, the asymptotic behavior for $R \to 0$ is not correct, since the penetration of charges has not yet been taken into account. An obvious problem is that, in the present form, $V_{\text{pol}}, V_{\text{ind}}$ and V_{disp} will become infinitely negative for $R = 0$. The problem arises because all power series were derived for the condition $R > (|r_A + r_B|)$, which is clearly not the case for small R values. Nevertheless, the functional form of (3.1) can be used if damping functions are introduced specifically to quench these terms at small R. Equation (3.1) is then rewritten as

$$V = A \exp(-bR)$$
$$- \sum_{n=3}^{\infty} f_n(R) \frac{C_{\text{pol},n}}{R^n} - \sum_{n=6}^{\infty} f_n(R) \frac{C_{\text{ind},n}}{R^n} - \sum_{n=6}^{\infty} f_n(R) \frac{C_{\text{disp},n}}{R^n} , \qquad (3.2)$$

with $f_n(R)$ representing the damping functions. The damping functions are constructed such that $f_n(R) \to 1$ for $R \to \infty$, since (3.1) is the correct asymptotic expansion. However, for $R \to 0$ the damping function $f_n(R)$ has to be such that $f_n(R)/R^n \to 0$ for $R \to 0$, to avoid the unphysical singularities in the expansion. Different types of damping functions have been used in the literature. Following the rigorous treatment of Mushed and Amos [138], which takes into account the corrections to conventional perturbation theory, it can be shown that the damping functions should be a product of a polynomial and an exponential. The most widely used are the Tang–Toennies damping functions [183], defined by

$$f_n(R) = 1 - \exp(-bR) \sum_{k=0}^{n} \frac{(bR)^k}{k!} .$$

The parameter b is the same as that appearing in the repulsive part of the potential. This function fulfills our requirements and, moreover, guarantees that the slope of the potential is finite at $R = 0$. It is physically reasonable that the parameter b appears both in the damping function and in the description of the repulsive part of the potential, since the two phenomena have the same physical origin, which is the overlap of charge distributions. Both damping and repulsion are assumed to fall off as $\exp(-bR)$ with increasing distance R. Other forms for the damping functions have been reported in the literature. For alkali mixtures, damping functions based upon the use of the ionization potentials of the monomers to fix the corresponding parameters have been constructed [43] .

The parameters $C_{\text{pol},n}$ can be calculated from monomer properties such as the multipole moments. However, the knowledge of $C_{\text{disp},6}$ requires the knowledge of frequency-dependent polarizabilities. These can be obtained either from semiempirical approaches [104] or from ab initio calculations [67]. For semiempirical studies of the intermolecular potentials, these parameters are either fitted to the repulsive part of the self-consistent field (SCF) ab initio potential surface or adapted to the experimental data. Determinations of potential surfaces from experimental data have so far been accomplished for the following complexes: Ar–HCl [90, 44], Ar–HF [93], Ar–HBr [91], Ar–H$_2$O [36], Ar–H$_2$ [12, 107], Ar–NH$_3$ [169], (NH$_3$)$_2$ [146], (HCl)$_2$ [47], Ar–C$_2$H$_2$ [8] and Ar–CO (Chap. 8). In all cases the functional forms of the intermolecular potentials were taken from (3.2). The fact that in all these studies the experimental data could be fitted within experimental uncertainty to (3.2) shows that it includes the terms most relevant to the description of intermolecular forces. However, the importance of terms describing charge transfer in a hydrogen-bonded complex is still not fully understood. The potential surfaces for two prototype systems, namely Ar–CO and $(NH_3)_2$, will be discussed in detail later.

3.2 Parameterization

Proceeding from the Hamiltonian introduced in (3.2), we restrict the discussion in this section to a single-center expansion in each part of the complex, since this is sufficient to describe Ar–CO. In order to fit the experimental data, a limited subset of the parameters can be varied. In general, the parameters introduced in this potential, A, b, C_{pol}, C_{ind}, C_{disp}, depend on the relative angular orientation of the monomers. In the case of a noble-gas–diatomic complex this orientation can be described by a single angle, θ, and the angular dependence of the potential can be written as a Legendre expansion:

$$V(R,\theta) = \sum_{l=0}^{\infty} V^l(R) P_l(\cos\theta) \, ,$$

with $P_0(\cos\theta) = 1, P_1(\cos\theta) = \cos\theta, \ldots$ providing an orthogonal basis set. The most obvious choice is, therefore, also to use a Legendre expansion for the parameters A, b, C_{pol}, C_{ind}, C_{disp}:

$$C_x = \sum_{l=0}^{\infty} C_x^l P_l(\cos\theta) \, .$$

The question that now needs to be addressed is which terms in the expansion need to be included to give a good overall description of the experimental data. One possibility would be to restrict the number of parameters by varying only the lowest-order terms of the Legendre expansion, e.g. C_x^0, C_x^1 and C_x^2, and set higher-order terms equal to zero. However, this is only a good choice if the expansion is rapidly convergent. If not, neglecting higher-order terms would give an unsatisfactory result. The following approaches are used in the literature to obtain an appropriate set of parameters.

One approach, the so-called Hartree–Fock plus damped dispersion (HFD) potential, was first introduced by Scoles and coworkers [1, 43, 44, 78].

The dependence of the repulsion was expressed as

$$A(\theta) R^{\gamma(\theta)} \exp[-\beta(\theta) R] \, ,$$

where the angular dependence of A, γ and β can be expressed by a Legendre expansion. These parameters were then obtained by fitting the short-range part of a high-level SCF ab initio calculation to the above expression. The SCF calculations neglect correlation effects (dispersion is not taken into account); however, they give rather accurate descriptions of the repulsive part of the potential. The results of such calculations include the induction energy, which can be subtracted before the fit is performed. This method was used for the prediction of the Ar–HCl, Ar–HF and Ar–C_2H_2 potential surfaces [8, 44].

The parameters $C^0_{\text{ind},6}$ and $C^2_{\text{ind},6}$ were calculated from the electrical properties of the monomers. $C^0_{\text{disp},6}$ was fixed by a value obtained from semiempirical calculations by Meath and coworkers (e.g. [104]). The higher-order terms $C^2_{\text{disp},6}$, $C^4_{\text{disp},6}$, $C^1_{\text{disp},7}$ and $C^3_{\text{disp},7}$ could then be obtained from a product of $C^0_{\text{disp},6}$ and monomer properties or, if available, also from ab initio calculations. In this way the parameters C^l_n were calculated by the use of scaling laws up to the required level.

A slightly different choice of parameters was used by Hutson [90] and was used for a fit of the potential surface to experimental data for the following complexes: Ar–HCl, Ar–HF, Ar–HBr, Ar–H_2, Ar–H_2O, Ar–NH_3. $C^0_{\text{ind},6}$, $C^2_{\text{ind},6}$, $C^2_{\text{disp},6}$, $C^4_{\text{disp},6}$, $C^1_{\text{disp},7}$ and $C^3_{\text{disp},7}$ were obtained in the same way as in the HFD model and were fixed in the course of the calculation. The parameters which were varied were the two parameters describing the repulsive part of the potential (A, b), and the higher-order dispersion terms $(C^0_{\text{disp},7}, C^0_{\text{disp},8}, \ldots)$. For very anisotropic potentials it is advantageous to express the free parameters A, b, $C^0_{\text{disp},7}$, $C^0_{\text{disp},8}$, $\ldots$ in terms of the well depth ϵ, the position of the minimum R_m, and b (e.g. [92]). These new parameters were expanded in a Legendre expansion and varied during the fit. This turns out generally to be a better choice, since ϵ and R_m are less correlated. Moreover, the Legendre expansions for these parameters converge faster, especially in the case of highly anisotropic potentials. This procedure has been used successfully for the following complexes: Ar–HCl, Ar–HF, Ar–HBr, Ar–H_2, Ar–H_2O and Ar–NH_3, as well as for a fit of the Ar–CO spectrum.

3.3 An Example: Ar–HF

In this section we consider the example of the potential surface of Ar–HF, which has been the subject of a large number of experimental and theoretical studies (see e.g. [93, 119] and references therein).

The most recent and accurate potential surface reported for Ar–HF is shown as a contour plot in two dimensions, namely R and θ, in Fig. 3.1. A repulsive wall, characterized by $\exp(-\beta R)$, is seen at small R, where there is steep rise in the potential surface. Two minima are also evident in the potential surface, at $\theta = 0°$ and $\theta = 180°$, which is surprising at first sight. R gives the distance between the argon atom and the center of mass of HF which is very close to the fluorine atom. Both minima are caused by the detailed balance of the different contributions to the intermolecular potential. The minimum at $\theta = 0°$ corresponds to the hydrogen pointing towards the argon, giving a maximum in the induction energy. HF is a strongly polar molecule

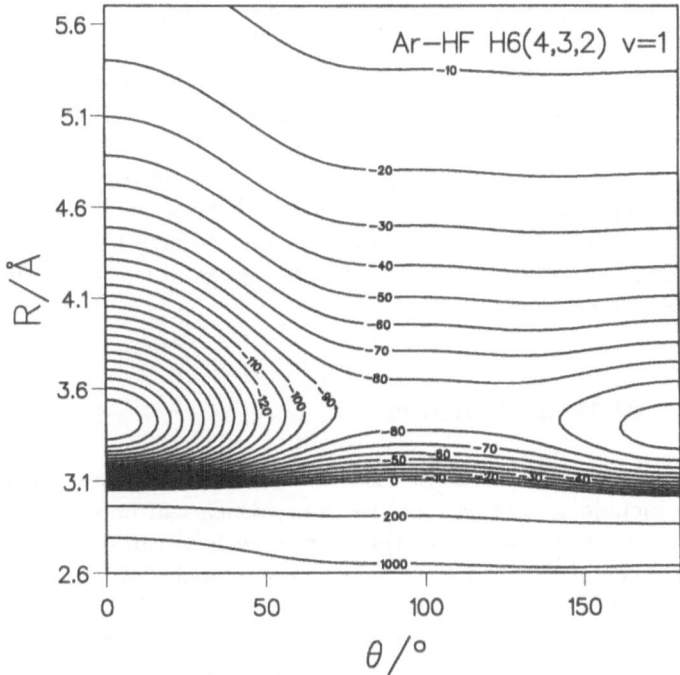

Fig. 3.1. The potential of Ar–HF, as obtained by Hutson

(dipole moment 1.8 D), so that the induction energy is therefore expected to make a significant contribution. The reason for the second minimum, at $\theta = 180°$, only becomes clear when we consider all of the contributions to the potential energy for different positions. These are listed in Table 3.1, where

Table 3.1. Contributions to the potential energy of Ar–HF

θ	Repulsion	Induction	Dispersion	R_m
0°	257 cm^{-1}	−72 cm^{-1}	−405 cm^{-1}	3.43 Å
90°	90 cm^{-1}	−8 cm^{-1}	−166 cm^{-1}	3.50 Å
180°	105 cm^{-1}	−6 cm^{-1}	−219 cm^{-1}	3.37 Å

R was chosen to correspond to the minimum of the potential surface for each angle θ, defining $R_\mathrm{m}(\theta)$, with R_m being the distance from the argon to the center of mass of the HF. We can see from Table 3.1 that the dispersion energy is the main contribution to the potential energy, even for this very polar molecule. The increase in the repulsion from 105 to 257 cm^{-1} when going from 180 to 0° can be explained in terms of the influence of the H atom. Since the H atom is pointing towards the argon atom at $\theta = 0°$, the distance between the H and the Ar atoms is only 2.56 Å, whereas for $\theta = 180°$ the argon–fluorine distance is 3.32 Å. The dispersion is approximately two times larger for $\theta = 180°$ than for $\theta = 0°$. In comparison, the repulsion is only slightly increased for $\theta = 180°$ compared with $\theta = 90°$, although the distance is considerably smaller. The latter is due to a lack of repulsion at $\theta = 180°$, which corresponds to a decrease in charge density as a result of the electron-withdrawing effect of the HF bond. We can see here that for an understanding of the potential energy surface of complexes all contributions have to be well determined.

3.4 Calculation of Bound States

The experimental quantities which can be measured to check the accuracy of a potential surface include scattering cross sections, which can be obtained from collision experiments (e.g. [168]), or the energy levels of van der Waals complexes, measurement of which is the subject of this book. We therefore have to introduce a method for the calculation of bound states using intermolecular potential surfaces. First, we have to find an adequate basis set and then solve the Schrödinger equation, using the potential of (3.2). The effect of the introduction of a weak potential will first be investigated qualitatively, to demonstrate the main features and differences compared with chemically bound molecules. The nuclear Hamiltonian of the complex is given by

$$H_0 = E_{\text{kin}} + V(R, \theta) \ .$$

In an initial simplification, we fix R at R_m and allow only variations in the angle θ. The potential can then be described by $V(\theta) = \sum_{l=0}^{\infty} V^l P_l(\cos \theta)$. We now want to study the effect of a simple, but representative potential. If we look at the potential for Ar–HF, as displayed in Fig. 3.1, we can see that, in a very crude approximation, the angular dependence can be described by $\cos^2 \theta$. We shall therefore, for a purely qualitative picture, restrict the potential to

$$V(\theta) = V^2 P_2(\cos \theta) = \frac{1}{2} V^2 (3 \cos^2 \theta - 1) \ .$$

The kinetic energy of a complex consisting of a noble gas and a diatomic can be described by the following Hamiltonian, if the kinetic energy of the center of mass and the vibrational energy of the monomer are separated:

$$H = b\tilde{j}^2 + B\tilde{l}^2 + V^2 P_2(\cos \theta), \tag{3.3}$$

where b and B are the rotational constants of the monomer and the complex, which are inverse proportional to the moment of inertia of the CO monomer and the Ar–CO complex, respectively. $\tilde{l}$ and $\tilde{j}$ are the operators for the angular momentum as defined in a space fixed axis system describing the rotation of the complex around its center of mass (called "end-over-end"rotation) and the rotation of the monomer ("internal rotation"). We shall choose as an example the Ar–CO complex since it will be the subject of studies later in this book. Since we want to study the influence of weak intermolecular forces, we shall choose a basis which is correct for $V^2 \to 0$. In this limit the CO monomer can perform an unhindered rotation, often referred to as the free-rotor limit. For highly anisotropic potentials (V^2 large), the complex will mimic a chemically bound molecule, this limit being called the rigid-rotor limit. Between these two limits, the complex is often described as a hindered rotor. In the free-rotor limit the quantum numbers j and l are good quantum numbers. If we increase the anisotropy, $\tilde{j}$ and $\tilde{l}$ will couple to $\tilde{J}$, the total angular momentum. In the free-rotor limit the basis functions $|jm_jlm_l\rangle$ will span an appropriate basis; in the rigid-rotor limit the basis functions $|JMK\rangle$, do so. Here K is the projection on an axis fixed in the molecule, whereas M, m_l, m_j are projections on an axis fixed in space. The hindered rotor is described in a basis using the basis functions $|JMjl\rangle$. The calculation of the energy levels for Ar–CO with an intermolecular potential of the form (3.3) requires a matrix H^{JM}, with elements

$$H_{jl;j'l'}^{JM} = \langle JMjl|H|JMj'l'\rangle ,$$

where j, l, j', l' fulfill the condition

$$|j - l| \leq J \leq j + l .$$

The eigenvalues of this matrix will give the energy levels in the hindered-rotor case. The matrix is given by

$$\begin{aligned} H_{jl;j'l'}^{JM} = \ & b_{CO}j(j + 1)\delta_{j,j'}\delta_{l,l'} + Bl(l + 1)\delta_{j,j'}\delta_{l,l'} \\ & + V^2\langle JM_Jjl|P_2(\cos\theta)|JM_Jj'l'\rangle . \end{aligned} \tag{3.4}$$

The corresponding eigenvalues give the energy levels in the hindered-rotor case. The lowest energy levels have been calculated for Ar–CO (with $b_{CO} = 1.93$ cm^{-1} and $B = 0.07$ cm^{-1}), taking V_2 to be positive, which corresponds to a T-shaped ($\theta = 90°$) minimum configuration, as has been observed for Ar–CO. The result, where V_2 is varied between 0 and 10 cm^{-1}, is shown in Fig. 3.2. For $V_2 = 0$ we see the free-rotor structure, with the different j levels separated by $2b_{CO}j$. On top of this rotational pattern there is the end-over-end rotation, which is described by $Bl(l + 1)$. If the anisotropy is not 0, the degeneracy for $j = 1$ is lifted and two stacks of levels separate with increasing V_2. One of these sets corresponds to an in-plane rotation, while the other is the out-of-plane rotation of the CO monomer. Although the out-of-plane rotation is relatively unhindered by the presence of an angular potential, the in-plane rotation of the CO monomer is hindered with increasing V_2 and

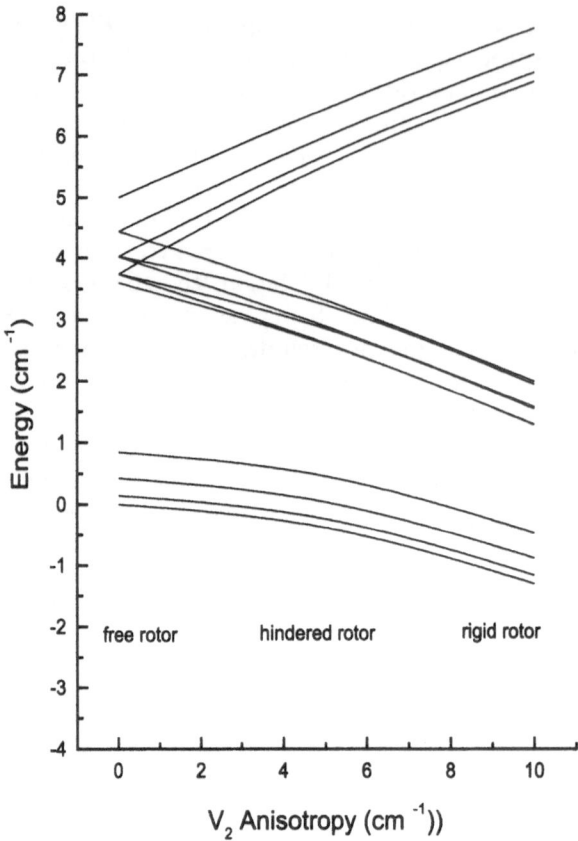

Fig. 3.2. Energy-level diagram for Ar–CO at $R = R_m$ as a function of the angular anisotropy V_2

hence increases in energy. As we approach the rigid-rotor limit the internal rotations of the CO monomer become strongly hindered and eventually become intermolecular bending vibrations, as in chemically bonded molecules. Again, the substructure corresponds to sequential J levels that describe the overall rotation of the complex. For large V_2 values the out-of-plane rotation corresponds to a rotation of the now rigid molecule with $K_a = 1$. Here K_a is the projection of the angular momentum on the intermolecular a-axis, corresponding to the axis which yields the largest moment of inertia (A). All levels are doubly degenerate and are separated by $AK_a^2 = A$ from the ground state. In the rigid-rotor limit, Ar–CO has a fixed T-shaped structure and A is equal to b_{CO}. According to quantum mechanics the energy for a three-dimensional rotation is $b_{CO}j(j+1) = 2b_{CO}$ (for $j=1$), whereas in the rigid-rotor limit the rotation is restricted to the plane containing the b and c axes of the complex. The energy is obtained as $AK_a^2 = A = b_{CO}$ (for $K_a = 1$), which is half the energy for the free-rotor limit. The anisotropy of the potential mixes levels with different values of j. The slight decrease of energy for the $j = 0$ levels is a consequence of this. Even though V_2 is only 10 cm^{-1} at the right-hand side of Fig. 3.2, the energy starts to resemble that of a rigid rotor: we can assign K_a quantum numbers and use J and K_a as quantum numbers, although we have to keep in mind that B and A do not have a pure structural meaning. In general, most complexes are intermediate cases (hindered rotor), in which an assignment of the data to J and K_a quantum numbers is possible. However, care has to be taken when structural implications are deduced.

In the general case of calculating bound energy levels in a potential $V(R, \theta)$, the basis is constructed as a product of a radial function and spherical harmonics, associated with the internal and the end-over-end rotation:

$$\psi^{JM}(R, \tilde{r}, \tilde{R}) = \chi_n(R) \sum_{m_l, m_j} \langle jm_j lm_l | JM \rangle Y^j_{m_j}(\tilde{r}) Y^l_{m_l}(\tilde{R}) \, ,$$

where χ_n are Morse-type functions or distributed Gaussians and $\tilde{r}, \tilde{R}$ describe the orientation of the intramolecular and intermolecular axis, respectively (see [193] for more details). If the angular potential is expanded in Legendre polynomials, the solution of the Schrödinger equation for this part can be directly calculated, as in the example above. However, calculations of bound states in a potential of the type $V(R, \theta)$ require an integration over the radial coordinate. Programme packages, such as BOUND by J. Hutson and ATDIAT, by Tennyson offer numerical solutions for this purpose.

3.5 Fitting the Potential Surface

If we want to fit semiempirical potential surfaces to the experimental observables, we have to use nonlinear fitting routines in order to adjust the parameters in (3.2) such that the measured energies for the bound levels are obtained. The fitting routine requires successive recalculation of the energies, which implies long calculation times. In order to speed up this process, two methods have been developed in which the time-consuming integration is replaced by the calculation of energies at discrete points. These two methods are the collocation method of Peet and Yang [152] and the discrete-variable representation (DVR) of Choi and Light [32]. Here, only the general ideas of these methods will be outlined. For a solution of the bound states, trial wave functions $|\psi_k\rangle$ are taken, which are expanded as linear combinations of basis functions $|\phi_j\rangle$. If we have

$$|\psi_k\rangle = \sum c_{jk} |\phi_j\rangle \, ,$$

we want to choose c_{jk} such that $(H - E_k)\psi_k$ is minimized. In variational methods this is achieved by requiring that

$$\sum_j \langle \phi_i | H - E_k | \phi_j \rangle c_{jk} = 0 \, , \; \forall i \, . \tag{3.5}$$

We can see from this equation that the calculation of the c_{jk} requires the calculation of multidimensional integrals, which is very time-consuming. Both of the above methods try to avoid these integration procedures, and replace them by pointwise calculations, as will be outlined. However, the main idea behind this is that the integrals are replaced by sums over values of the functional at discrete, carefully selected points. In the collocation method this is accomplished in the following way. Instead of requiring that $(H - E_K)\psi_k$ is zero everywhere, we require that this is true only at certain points r_i, $i = 1, \ldots n$, called the collocation points. We thereby obtain the equation

$$\langle r_i | (H - E_k) | \psi_k \rangle = 0 \, ,$$
$$\sum_j \langle r_i | (H - E_k | \phi_j \rangle c_{jk} = 0 \, , \; \forall i \, . \tag{3.6}$$

This replaces the previous (3.5). The solution requires the diagonalization of a nonsymmetric $N \times M$ matrix, where N is the number of collocation points and M the number of basis functions. In contrast to the variational method (3.5), no integrals have to be calculated. The collocation points r_i have to be carefully selected such that the approximation of the integral is accurate. For specific basis functions, specific schemes exist in the literature for doing this (e.g. quadrature schemes).

The DVR method uses the same principle: replacement of integrals by sums. However, in the DVR method, only the potential-energy matrix is approximated by such quadrature schemes, not the complete Hamiltonian. The method uses two basis sets: $|\varphi_n\rangle$ and $|\phi_n\rangle$. $|\varphi\rangle$ is an analytical basis function. Analytical basis functions lead to simple expressions for the kinetic-energy operator if the appropriate basis set is used. In the example above, we could simply use the expression

$$Bl(l+1) + b_{\mathrm{CO}} j(j+1) \, ,$$

which is already diagonal. However, a pointwise representation is used for the potential-energy operator. The transformation from one basis to the other is given by

$$\phi_k(x) = \sum_n \varphi(x) \varphi_n(x_k) \sqrt{w_k} \, ,$$

with w_k being the weight. The potential energy in the analytical basis is approximated by a sum over pointwise-calculated functional values times the weights, i.e.

$$\langle \varphi_n | V | \varphi_m \rangle = \sum_k w_k \varphi_n(x_k) V(x_k) \varphi_m(x_k) \, .$$

Once again, the points x_i and weights w_i have to be carefully selected. It can be demonstrated that the kinetic energy is diagonal in the new basis ϕ_k. In practice, the kinetic-energy part is calculated in the ϕ_k basis; the potential energy is calculated in the φ_n basis and afterwards transformed to the ϕ_k basis. Using this method, the advantages of both methods are exploited. The calculation requires only the diagonalization of a symmetric matrix, which speeds up the calculation time compared with the collocation method. On the other hand, the collocation method is easier to handle and to program. However, the fastest routines currently used are based on the DVR approach (see [101] for a comparison). The disadvantage compared with variational methods is that the eigenvalues are no longer upper bounds of the exact eigenvalues, but are only approximate. However, in practice these techniques have turned out to provide fast and sufficiently accurate solutions (see also [8, 36]).

In the following chapters we give an introduction to the spectroscopic experimental techniques used for studying intermolecular forces.

4. The Molecular Beam

4.1 The Supersonic Expansion

The investigation of small molecular complexes has recently become an important source of information on intermolecular forces. These small systems are at the limit of what can be accurately described by theoretical (ab initio) models. A description of larger systems, such as surface interactions or fluids, is beyond the scope of what can be achieved at the moment. Molecular complexes have therefore turned out to be small, theoretically tractable systems, which are excellently suited for the investigation of intermolecular forces. The experimental techniques which are used for the study of molecular complexes are scattering experiments and spectroscopy. In both cases, efficient production methods are required. Since the van der Waals interaction is quite weak, most complexes are not bound at room temperature. An efficient production method therefore requires a decrease in temperature in order to achieve a thermal energy which is less than the binding energy (typically 150 K). Even in the case of strongly bound complexes, such as hydrogen-bonded complexes, a decrease in temperature will increase the number of dimers compared with monomers. In order to produce very low temperatures, one can either cool the gas in a cell surrounded by a liquid at low temperature (e.g. liquid nitrogen) or use molecular beams. Since the molecular-beam technique is the method used in our laboratory and by most other groups, the discussion will be restricted to the production of complexes in molecular beams. A very detailed introduction to molecular beams is given in [173]. A short introduction is given here, which summarizes Chap. 2 of [173] and an article by Hagena about the nucleation and growth of clusters in expanding nozzle flows [68].

In a molecular beam, the gas is expanded through a nozzle into a vacuum. The situation is sketched in Fig. 4.1. The gas starts from a temperature T_0 and a pressure P_0. The chamber is evacuated by a pump (a mechanical pump or an oil diffusion pump) and is kept at a background pressure P_b. The gas is accelerated by the pressure difference $P_0 - P_b$. As the mean velocity increases, the enthalpy of the gas decreases and the gas is thus cooled down to a very low temperature. If the pressure ratio P_0/P_b is larger than 2, we have a supersonic expansion, which means that the Mach number M, which is equal to the ratio of the mean velocity of the gas to the speed of sound, is larger than 1. In a supersonic expansion, the gas leaves the nozzle with

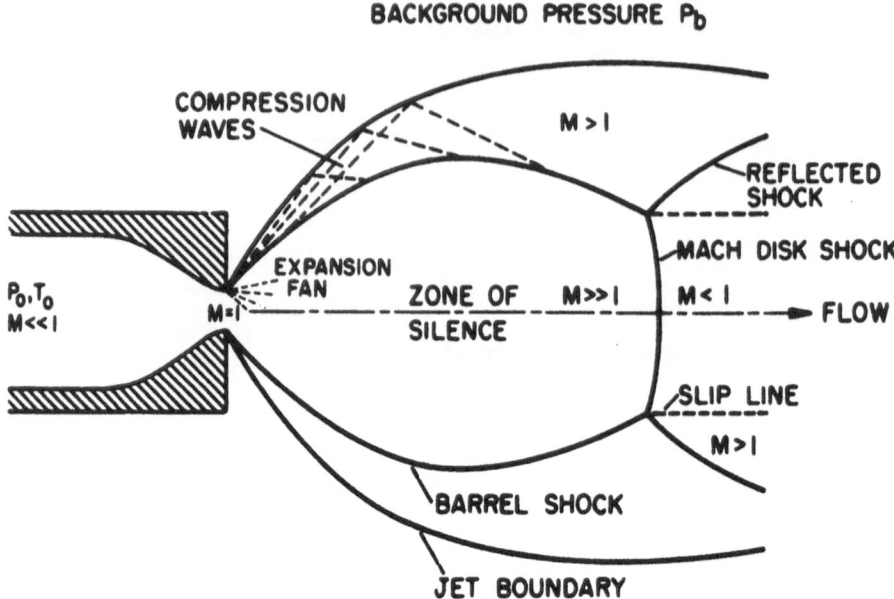

BACKGROUND PRESSURE P_b

COMPRESSION
WAVES

$M > 1$

REFLECTED
SHOCK

MACH DISK SHOCK

P_0, T_0
$M \ll 1$

EXPANSION
FAN

$M = 1$

ZONE OF $M \gg 1$ $M < 1$
SILENCE

FLOW

SLIP LINE

$M > 1$

BARREL SHOCK

JET BOUNDARY

Fig. 4.1. Molecular-beam expansion [173]

a pressure which is larger than the chamber pressure, and a subsequent expansion occurs until the flow meets the boundary condition imposed by the background pressure P_b. The gas overexpands, since the information about the boundary conditions can only be transported at the speed of sound. The gas is then compressed by shock waves, which mean regions of large density, pressure, temperature and velocity gradients, in order to meet the boundary condition. The expansion can be characterized by the parameters P, T, n and v, which are the pressure, temperature, density and velocity, respectively. These quantities can be calculated if we assume the expansion can be treated as that of an ideal gas. The acceleration of the gas is connected to a decrease in temperature. We can calculate the flow velocity v as

$$v^2 = 2 \int_T^{T_0} C_P \, dT = 2C_P(T_0 - T) \,. \tag{4.1}$$

The maximum flow velocity which can be reached is therefore given by

$$v_{\max} = \sqrt{2C_P T_0} = \sqrt{\frac{2R}{W} \frac{\gamma}{\gamma - 1} T_0} \,,$$

where C_P, R, W and γ are the heat capacity, the gas constant, the molecular weight and the ratio of the specific heats. This velocity can be compared with two typical velocities:

1. The thermal velocity before the expansion, which is:

$$v_{\text{therm}} = \sqrt{\frac{2RT}{W}} \ .$$

2. The speed of sound, which is

$$a = \sqrt{\frac{\gamma RT}{W}} \ .$$

If we compare the different velocities, we see that the maximum velocity for monatomic gases ($\gamma = 5/3$) is approximately 1.6 times greater than the mean thermal velocity before the expansion (at $T = T_0$). However, the speed of sound decreases very quickly in an expansion, owing to the decrease in temperature. We therefore find large Mach numbers, meaning large ratios $M = v/a$, in the expansion. The expansion becomes a supersonic expansion. The velocity is given by $u = Ma = M\sqrt{\gamma RT/W}$.

The thermodynamic laws yield the following relations for ideal gases:

$$T^\gamma P^{1-\gamma} = \text{const.} , \tag{4.2}$$

$$P = nRT , \tag{4.3}$$

with T being the temperature and P the pressure of the gas. If we know the Mach number, we can calculate the velocity as a function of the temperature. If we insert this in (4.1), we can relate the temperature to the Mach number. Using this equation, we can calculate the dependence of the parameters P and n on the Mach number M:

$$\frac{T}{T_0} = \left(1 + \frac{\gamma - 1}{2} M^2\right)^{-1} , \tag{4.4}$$

$$v = M\sqrt{\gamma \frac{RT_0}{W}} \left(1 + \frac{\gamma - 1}{2} M^2\right)^{-1/2} , \tag{4.5}$$

$$\frac{P}{P_0} = \left(\frac{T}{T_0}\right)^{\gamma/(\gamma-1)} , \tag{4.6}$$

$$\frac{n}{n_0} = \left(\frac{T}{T_0}\right)^{1/(\gamma-1)} = \left(1 + \frac{\gamma - 1}{2} M^2\right)^{-1/(\gamma-1)} . \tag{4.7}$$

The Mach number for various gases as a function of x/d, where x is the distance from the nozzle and d the nozzle diameter, can be found in, for example, [173]. However, we can see from the equations that the flow velocity will increase very rapidly with M and will then quickly approach a constant value equal to $v_{\max}$. In contrast, the other parameters P, T and N would, under ideal conditions, continue to decrease with increasing Mach number. However, this decrease is limited by the appearance of shock waves, which terminate the expansion. For the temperature, we have a second limiting condition: the temperature only decreases as long as there are enough collisions to establish a thermal equilibrium. As soon as the number of collisions becomes smaller than one, the temperature in the expansion will not decrease any more, although the Mach number can still increase. This phenomenon is described by the quitting-surface model. In the collisionless regime, the molecules have constant temperatures $T_\parallel$, with $T_\parallel$ describing the spread in velocities parallel to the direction of the expansion. In the subsequent two sections we shall discuss the characteristics of the expansion for different nozzle types.

4.2 Axisymmetric Versus Planar Expansion

We shall now specify certain nozzle types and investigate the differences in the expansion. We first consider an axisymmetric expansion resulting from a point nozzle. Typical nozzle diameters are in the range 30–100 μm. A few nozzle diameters away from the nozzle, the streamlines are almost straight and the flow velocity approaches its final value. The density decreases with increasing distance from the source as x^{-2}. The density in the expansion is, further, proportional to d^2. With $\delta = x/d$, we obtain

$$\frac{n}{n_0} = K d^2 x^{-2} = K \delta^{-2} \ . \tag{4.8}$$

The constant K is equal to 0.15 or 0.086, depending on whether a monatomic gas ($\gamma = 5/3$) or a diatomic gas ($\gamma = 7/5$) is expanded [68].

A planar expansion is an expansion from a slit nozzle. Ideally, the slit would have a certain width d and would be infinitely long. However, in practice the length is limited to a certain value L. The slit nozzle used in our setup is shown in Fig. 4.2. The slit length is 5 cm and the width is typically 50–70 μm. For an ideal slit nozzle, the gas can only expand in one dimension, such that the density decreases as x^{-1}. We can, further, state that the density in the expansion is proportional to n_0 and to the slit width d. In general, we can summarize the dependence of n/n_0 on δ as

$$\frac{n}{n_0} = K d^i x^{-i} = K \delta^{-i} \ , \tag{4.9}$$

with $i = 1$ for a planar expansion and $i = 2$ for an axisymmetric expansion. For a planar expansion, the constant K is 0.2 for a monatomic gas and 0.136 for a diatomic gas [68]. Using the thermodynamic calculations given in the previous section, we can relate the quantities P, T and M to the density and therefore obtain the dependence of these quantities on the reduced source distance δ:

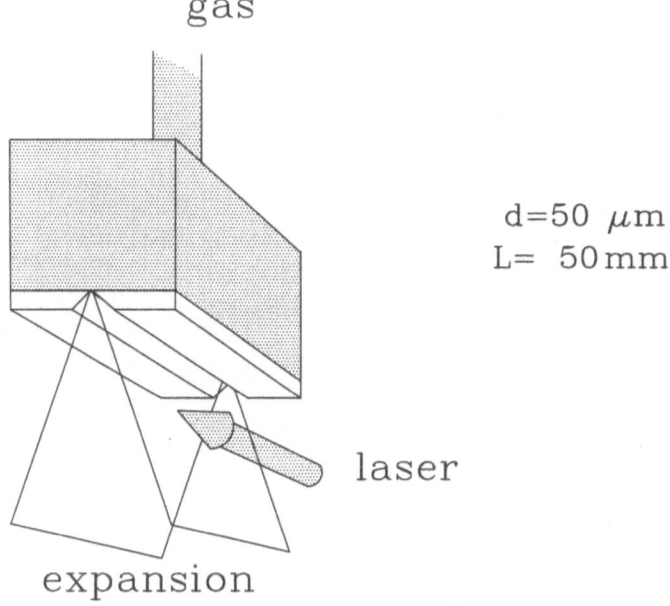

gas

d=50 μm
L= 50 mm

laser

expansion

Fig. 4.2. Slit nozzle used in infrared and far infrared spectroscopy of cluster

$$\frac{T}{T_0} = \left(\frac{n}{n_0}\right)^{\gamma-1} = K^{\gamma-1}\delta^{-i(\gamma-1)} , \qquad (4.10)$$

$$\frac{P}{P_0} = \left(\frac{T}{T_0}\right)^{\gamma/(\gamma-1)} = K^{\gamma}\delta^{-i\gamma} . \qquad (4.11)$$

If we approximate (4.4) for large Mach numbers and write

$$\frac{T_0}{T} = 1 + (\gamma-1)\frac{M^2}{2} \approx (\gamma-1)\frac{M^2}{2} ,$$

we can express the Mach number as a function of T/T_0 and therefore as a function of δ. If we insert (4.10) we obtain

$$M = \left(\frac{\gamma-1}{2}\right)^{-1/2} K^{-(\gamma-1)/2} \delta^{i/2(\gamma-1)} . \qquad (4.12)$$

It is interesting to compare the results for a pinhole and a slit nozzle. The characteristic properties of the two different types of expansion are given in Figs. 4.3 and 4.4. If we have a point nozzle and a slit nozzle, both with $d = 100$ μm and a pressure P_0 of ca. 100 kPa , the pressure in the point nozzle expansion will decrease to 1.7 Pa at $x = 1$ mm and to 8×10^{-4} Pa at $x = 1$ cm. In comparison, the pressure in a slit nozzle expansion at 1 mm will still be 130 Pa, and at 1 cm it will amount to 3 Pa. The Mach numbers

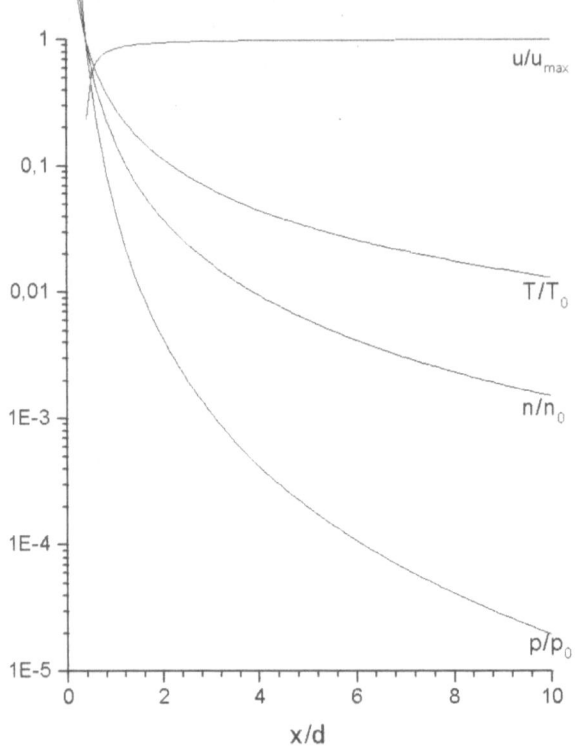

Fig. 4.3. Characteristic properties of an axisymmetric expansion ($d = 50$ μm) versus distance in units of source diameter

reached at 1 cm are 14 for the planar expansion and 70 for the axisymmetric expansion.

For an axisymmetric expansion, the Mach disk occurs at a distance of

$$x_M = 0.67 \, d \sqrt{\frac{P_0}{P_b}} \, ,$$

where P_b being the background pressure [173]. However, for a planar expansion this relation, which is based on empirical studies, could not be confirmed experimentally. Using a discharge nozzle, as described in Sect. 4.4, the Mach disk could be clearly observed at a distance of 2.5–3.5 cm for $d = 50$ μm and $P_0 = 70$–15 kPa. The dependence on P_0 and P_b could be better described as $x_M \propto P_0^{1/9}/P_b^{1/3}$; however, no final relation can be given yet, since only a few gases have been studied so far. The number of collisions in an expansion is especially important for the production of complexes. This issue will therefore be discussed in more detail in the next section.

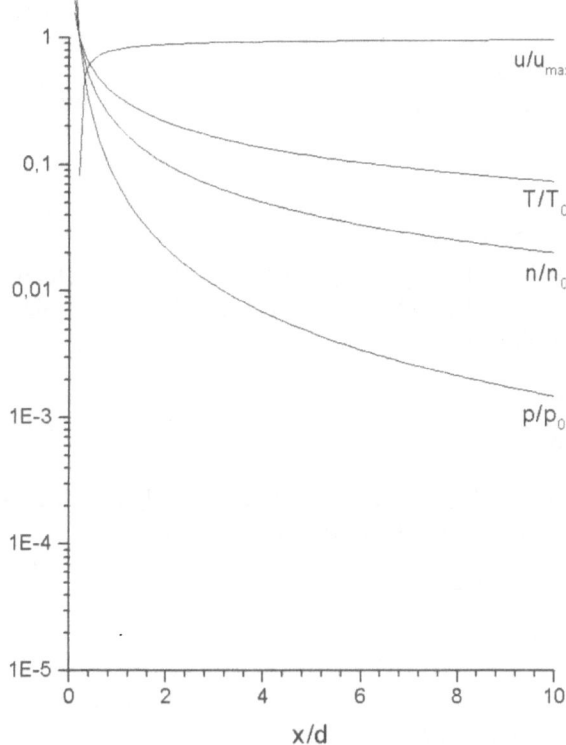

Fig. 4.4. Characteristic properties of a planar expansion ($d = 50$ μm) versus distance in units of source diameter

4.3 Generation of Clusters

The gas expands along the isentrope until it reaches a point where the gas is highly supersaturated. The exact onset of condensation is difficult to predict and depends on many parameters, such as the thermodynamics, the kinetics and the timescale of the expansion. The kinetics include the exchange of mass and energy due to monomer–cluster and cluster–cluster collisions and to spontaneous evaporation.

The formation of a cluster is described in two steps:

$$A + A \rightarrow A_2^* \, ,$$

$$A_2^* + M \rightarrow A_2 + M^* \, .$$

By the release of binding energy, the dimer is formed in an excited state. The collision with a third body M is therefore necessary for the stabilization of the complex. We can see that the formation of cluster is limited by the number of collisions. Although it is impossible to predict exactly the cluster formation rate in a mixture of gases, it is possible to predict the effects of changes of T_0, P_0 and d on the cluster formation.

The number of two-body collisions in a gas in the temperature interval $[T, T+dT]$ is given by

$$dZ \propto \sigma n v_{\text{therm}} \frac{dt}{dT} \, dT \, , \tag{4.13}$$

where σ is the cross section, n the density, v_{therm} the local thermal velocity and dt the transit time for passing through the interval $[T, dT]$. The cooling rate dT/dt is calculated according to

$$\frac{dT}{dt} = v \frac{dT}{dx} = \frac{v \, dt}{d \, d\delta} \propto \frac{v}{d} \frac{T}{\delta} = \frac{v}{d} T \left(\frac{T}{T_0} \right)^{1/(i(\gamma-1))} \, , \tag{4.14}$$

where dT/dx is obtained from (4.10). We want to obtain the explicit dependence of dT/dt on T_0, and therefore use the following relation:

$$v = v_{\text{therm}} \left(\frac{\gamma}{\gamma-1} \right)^{1/2} \left(\frac{T}{T_0} \right)^{-1/2} \, .$$

Inserting this, we obtain

$$\frac{dT}{dt} \propto \frac{v_{\text{therm}}}{d} T \left(\frac{T}{T_0} \right)^{1/(i(\gamma-1))} \left(\frac{T}{T_0} \right)^{-1/2} \, . \tag{4.15}$$

Combining (4.7), (4.13) and (4.15), we obtain

$$dZ \propto \sigma n_0 d \frac{dT}{T} \left(\frac{T}{T_0} \right)^{(i-1)/i(\gamma-1)} \left(\frac{T}{T_0} \right)^{1/2} \, . \tag{4.16}$$

The time dt for which T is in the interval from T to $T+dT$ is given by

$$dt \propto \frac{dT}{T} \left(\frac{T}{T_0} \right)^{-1/i(\gamma-1)} \left(\frac{T}{T_0} \right)^{1/2} \, . \tag{4.17}$$

In order to illustrate these results, the example of a planar expansion ($i = 1$) with a monatomic gas ($\gamma = 5/3$) has been chosen. For a given cluster, the rate of formation will depend on the density and the transit time dt. A higher cluster density and a less rapid expansion will favor the formation of a cluster. If we examine the influence of the temperature T_0, we can state that under otherwise unchanged conditions the density n will increase with decreasing T_0, yielding an increase in cluster formation. However, at the same time, the time dt for cluster formation will be reduced. We can, further, investigate the influence of the temperature T_0 on the collision rate dZ. From (4.16) we can deduce that the number of collisions increases with decreasing T_0. Collisions are responsible for the formation and decay of the clusters; however, for small clusters the first effect dominates. This means that the number of small clusters increases with increasing collision rates. In general, cluster formation will depend in a complex way on the density and collision rates. Qualitatively, we expect an increase of the cluster density with decreasing temperature T_0 owing to the increase of dZ and n.

As an example, we shall choose the case of a planar expansion of a monatomic gas. If we decrease the temperature T_0 from 300 to 80 K and keep all other parameters such as n_0 and d fixed, we find, according to (4.16), an increase in the number of collisions by a factor of two. If we look at the change of the density n, we obtain $n \propto (T_0)^{-1/(\gamma-1)}$ and therefore an increase in the density by a factor of seven. However, at the same time, the time for cluster formation is reduced, according to $dt \propto T_0^{1/(\gamma-1)-1/2}$, which means it is decreased by a factor of four, which leads to a less rapid cluster formation. Considering all three changes, we expect an increase in cluster formation if T_0 is decreased from 300 to 80 K. As a consequence, the temperature in the jet will be increased since, by formation of clusters, the binding energy is released as additional energy in the expansion. Experimentally, we were able to observe an increase in the final rotational temperature of the Ar–CO clusters (from 11 to 16 K) with decreasing nozzle temperature T_0 (from 300 to 80 K), which confirms our expectations.

We can, further, compare the two types of expansion (axisymmetric versus planar expansion) if we consider the number of collisions per interval dx. Using $dZ/dx = dZ/dT \, dT/dx$ with (4.10), we obtain for the axisymmetric expansion

$$\frac{dZ}{dx} \propto n_0 \delta^{-\gamma} . \tag{4.18}$$

For a planar expansion the corresponding equation is

$$\frac{dZ}{dx} \propto n_0 \delta^{-(\gamma-1)/2}. \tag{4.19}$$

We can see, by comparison of (4.18) and (4.19), that the number of collisions falls off much more rapidly in an axisymmetric expansion than in a planar expansion. In an axisymmetric expansion the number of (two-body) collisions is typically on the order of one for $\delta \geq 10$. Beyond this distance we shall have no more cooling in the expansion and the temperature in the expansion will typically freeze at 2–3 K. (If lower temperatures are required, T_0 can be reduced, or P_0 can be increased.) For a planar expansion the number of collisions decreases much more slowly. For $\delta \geq 10$, typically 100 collisions will still take place. However, the cooling is much slower, such that the expected temperature would be typically in the range of 7–8 K when the shock waves were reached. In reality, owing to effective cluster formation, temperatures of 11–16 K are reached.

Experimentally, we find for a planar expansion ($d = 100$ μm, $L = 5$ cm) that, after several cm ($\delta \geq 100$), the number of collisions is still high enough to establish a Boltzmann distribution. As a consequence, double resonance experiments lack a state-specific depopulation, since the high number of collisions guarantees a redistribution of populations. Whereas double resonance experiments are a good tool for the assignment of states for axisymmetric expansions, for planar expansions double resonance can only be used to distinguish between different spin states, since the nuclear spin is a preserved quantity in collisions.

4.4 The Discharge Nozzle

After the preceding general introduction to supersonic expansions, we now want to describe a specific nozzle design which was developed in Bonn in cooperation with the laser and molecular physics group in Nijmegen [83]. The experimental goal was to develop an efficient production method for radical complexes. For all radicals (with the exception of NO and O_2) we need an additional tool for the production of these short-lived species. One approach is to use a gas discharge for this purpose. In order to form radical complexes, a gas discharge to generate the radicals and a supersonic expansion to cool down the gas for cluster formation have to be combined. In 1983 the development of a corona-excited supersonic expansion using a point nozzle was reported by Droege and Engelking [45]. These authors designed the discharge nozzle so that the electric discharge is constrained to pass axially through the throat of an insulated nozzle. This is realized by using a glass tube with an opening of typically 50–200 μm as a nozzle. The electrode material is tungsten, situated just behind the nozzle on the high-pressure side. The discharge extends from the high-pressure to the low-pressure side, where it is grounded. The excitation of molecules by the discharge causes a heating of the gas. The temperature in the discharge nozzle therefore do not reach the low values achieved in normal supersonic expansions. The density and the current density fall off as $1/r^2$. The rate of excitation is roughly proportional to the product of both, falling off as $1/r^4$. The generation of radicals or ions occurs, therefore, only in the throat of the nozzle or within the first few nozzle diameters. Afterwards, the ions participate in the expansion, cooling down to lower temperatures. The fact that the excitation is limited to a small space allows the production of cold ions or even ion clusters. In direct absorption spectroscopy this nozzle design provides only a small absorption length, which is restricted here to several nozzle diameters, and low radical densities. In order to extend the absorption length, much work was undertaken to construct a discharge slit nozzle. The inherent difficulty of such a design is the extension of the discharge over the whole slit length. The first successful attempts to generate cold radicals used pulsed slit jets. The radical production was accomplished by excimer laser photolysis [35, 39]. Comer and Foster reported in 1993 the use of a continuous corona-excited slit nozzle expansion to generate CS and NH_2 radicals [37]. In order to stabilize the discharge over the whole length (5 cm), a new type of discharge nozzle has been constructed, which is shown in Fig. 4.5. A discharge will normally run on a single spot, where the highest E-field is locally reached. We therefore used several independent electrodes. Each of them is connected to the high-voltage power supply through a resistance of 100 kΩ.

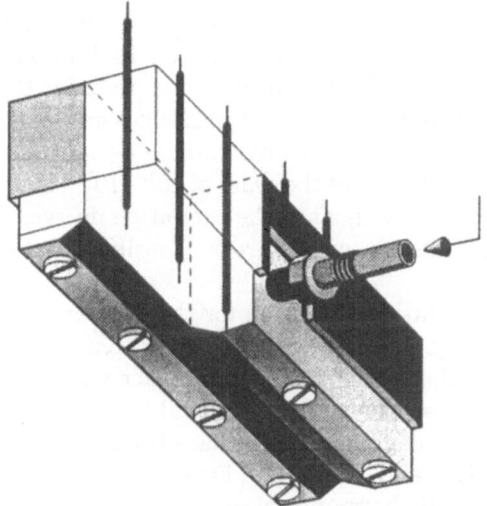

Fig. 4.5. Setup of discharge slit nozzle

Since each of the electrodes is in parallel sequence, even if one of the electrodes does not light, there is a voltage drop between the electrode and ground. The voltage drop corresponds to the overall voltage drop, since the resistance in the case when no discharge is lighted is infinite compared to 100 kΩ. If the discharge is lighted the voltage drop between the electrode and ground decreases to the burning voltage. If the voltage drop exceeds the starting voltage, the discharge will light again at that electrode. The resistances were selected so that this effect can be used; however, the voltage drop along each resistance must not be too high, because otherwise the voltage drop along the discharge might fall below the necessary burning value. If the resistance was absent, the discharge would light at one single point. The voltage drop along the other tips would then correspond to the burning voltage, which is less than the starting voltage. A stable discharge along a longer path can therefore only be established with the use of separated electrodes as described here. A discharge design with a single corona pin behind the slit nozzle for the measurement of ions has been reported by Xu et al. [201].

In our setup we used five to seven electrodes of tungsten, which were covered with glass for electrical insulation, except for a small, sharpened tip. The slit has the dimensions 5 cm × 150 μm. The backing pressure was 40 kPa and the chamber pressure about 30 Pa. Typical discharge conditions were 5 kV at the high-voltage side. The current was in the range 50–80 mA. Using this setup, the rotational spectrum of the N_2H^+ ion could be detected in the far-infrared (FIR) region [83]. The rotational temperature was determined to be 12 K. This demonstrated that in our setup high densities of cold ions or radicals could be produced, allowing for direct absorption spectroscopy. If we compare this design with the Droege–Engelking design, we can state that in our expansion the density and current density fall off as $1/r$. The excitation will therefore decrease as $1/r^2$, which is much slower than for the point nozzle. The density of radicals generated in the slit nozzle design should therefore exceed by far the density obtained in a point nozzle design. The density of generated radicals is estimated to be of the order of 10^{10}–10^{11} cm^{-2}. However, considerable cooling is still achieved, since the excitation decreases more rapidly than the number of collisions which are responsible for the cooling.

In the meantime, other designs of slit-jet expansion for the generation of radicals or ions have been reported (e.g. [69, 163]). Further improvement has been achieved by H. Linnartz in Basel, who used electron impact ionization of gas that is expanded through a slit nozzle [113]. Rotational temperatures as low as 15 K and densities of 10^9–10^{10} complexes cm^{-2} have been obtained. Examples are N_4^+, N_2–H^+–N_2, Ar–HN_2^+ and Ar–HCO^+. Further progress has been achieved with pulsed supersonic discharge plasmas [2, 98].

5. Infrared Spectroscopy of Complexes

5.1 Overview

Overviews of infrared laser spectroscopy of molecular complexes have been given by Miller [133], and by Leopold et al. in a review of current themes in microwave and infrared spectroscopy of weakly bound complexes [108]. I shall therefore give only a short summary here. Infrared spectroscopy is a well-established technique for the investigation of stable molecules. Infrared spectroscopy started in the 1930s (see [82]) and yielded a wealth of information on the structure and dynamics of chemically bound molecules. Infrared spectroscopy has been extended to the study of molecular complexes ever since laser techniques were combined with molecular-beam techniques. Early infrared studies of molecular complexes involved the use of long-path cells, which were cooled to low temperatures [154]. The first rotationally resolved spectroscopic studies of molecular complexes in molecular beams were performed in 1986–1988 [22, 132, 144, 155, 157]. Three distinct techniques are currently in use. The guiding principle of each technique will be presented. Detailed descriptions of the infrared laser systems involved can be found in the references below, and a more general description is given in [41] by Demtröder.

5.2 Direct Absorption Spectroscopy

The laser beam passes through the molecular beam, and the absorption of the laser beam is measured as a function of frequency. The signal, which corresponds to the decrease of laser beam intensity, is given by $\delta I = I_0 \exp -\alpha L$, where I_0 is the laser intensity, α the absorption coefficient and L the path length. The sensitivity of this method can be limited either by detector noise or by amplitude fluctuations of the laser. If the laser amplitude noise exceeds the detector noise, as in most cases, the expected signal-to-noise ratio is independent of I_0, but increases linearly with α and L, as long as $\alpha L \gg 1$. As a consequence, techniques have been developed to increase α by an increase in the gas density. This was first accomplished by the introduction of pulsed nozzles [75, 135], providing for short times nozzle densities which are orders of magnitude higher than in continuous expansions. A further improvement over point nozzle sources was achieved by the use of slit nozzle expansions [117]. As described in the previous chapter, this type of expansion provides high gas densities. In addition, cluster formation is enhanced by an increased number of collisions. The laser beam crosses the free expansion, which is several centimeters long. This implies an enormous increase in the absorption length L compared with that of a few millimeters for point nozzle expansions. The Doppler width in slit expansions is decreased by a factor of 5–7 compared with room temperature. This is less than for a skimmed point nozzle expansion, but turns out to be sufficient in several cases, especially if the laser line width is comparable to the observed Doppler width. The absorption path can be further increased by the use of multipass cells. For an increase in sensitivity, phase-sensitive detection methods such as frequency modulation or concentration modulation by pulsed nozzles (e.g. [42]) have been applied. As continuous laser sources, the following infrared lasers have been used:

1. The difference frequency laser, developed by Pine [154]. Developments of this technology are described in [156, 177]. Lasers are available with frequencies in the range 2400–4550 cm^{-1}; typical output powers are on the order of 1 µW.
2. Diode lasers. This type of laser will be described in more detail in Sect. 6.2. Lead salt diode lasers with an output power of up to 1 mW are available in the frequency range 650–3300 cm^{-1}.
3. Color center lasers. Typical laser output powers are 10 mW in the range 3000–4500 cm^{-1} and 100 mW in the range 6000–7000 cm^{-1}.
4. Tunable far-infrared lasers. This type of laser was developed in Nijmegen in 1978 [11]. The sideband laser can be operated from 500 GHz up to 3000 GHz with a typical output power of several 100 µW. The same type of laser was used in the studies of water complexes in Berkley [13].
5. Optical parametric oscillators (OPOs). An new design for a continuous tunable IR laser involves the use of an optical parametric oscillator. Schneider et al. [171] reported the operation of a quasi-phase-matched single-resonance OPO widely tunable in the ranges 1.66–1.99 µm and 2.29–2.96 µm, with an output power of up to 50 mW. Applications have been reported by Kühnemann et al. [102] and Gibson et al. [59].

Another method, which is complementary to laser spectroscopy is Fourier transform infrared spectroscopy in supersonic free-jet expansions, which has been used by some groups [65, 120, 127]. This has the advantage of a fast broadband access to the spectrum; however, it lacks the resolution achieved in laser absorption experiments.

5.3 Optothermal Detection

This technique makes use of the fact that thermal detectors, in particular cryogenic bolometers, can be used as beam detectors. A bolometer has an electrical resistance which depends strongly on the temperature. It can therefore detect small temperature changes ($\delta T = 3 \times 10^{-8}$ K) very efficiently. The sensitivity of this device increases as the bolometer is cooled down to very low temperatures, e.g. to 4 K (liquid helium) or even less. Temperature changes are caused by the energy of the beam, which is released at the bolometer surface. The first application as a beam detector in scattering experiments was reported by Cavallini et al. [27]. Spectroscopy using optothermal detectors was pioneered by Gough, Miller and Scoles [61]. A diagram showing a schematic setup for optothermal spectroscopy is shown in Fig. 5.3. A point nozzle expansion is used for the generation of molecu-

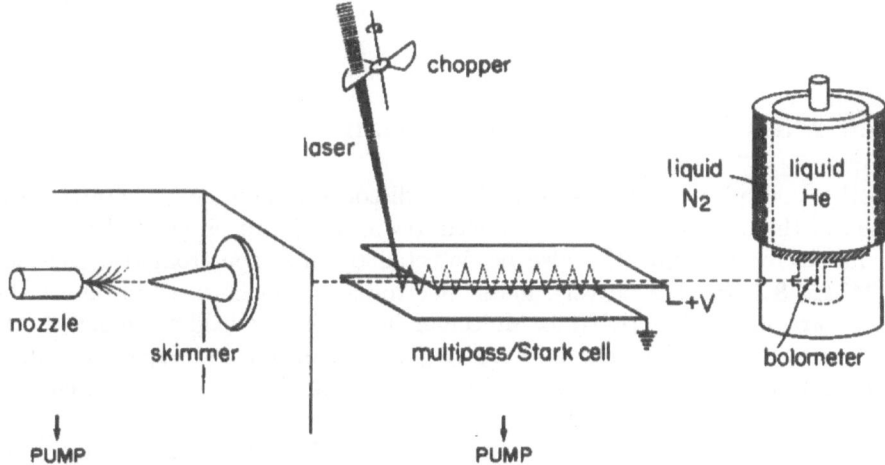

Fig. 5.1. Experimental setup for optothermal detection [173]

lar complexes. The molecular beam is extracted via a skimmer into a second chamber. The skimmer is used for collimation [41] of the beam and makes possible a differential-pumping scheme, thus decreasing the background pressure to a very low value. The molecules in the beam can be excited by a modulated infrared laser beam, which crosses the molecular beam. The molecules do not undergo further collisions at that point. The molecules or molecular complexes can therefore carry additional energy to the bolometer, which is detected as a modulated increase in temperature. In absence of the infrared laser source, the bolometer will experience a small, unmodulated temperature shift due to the translational and adsorption energy of the molecular beam. If the infrared laser radiation leads to dissociation of the molecular complex on a timescale which is less than the flight time to the bolometer, the absorption of light can be detected as a loss of energy at the detector. Examples of the application of this technique to the investigation of molecular complexes are given in [62, 132, 178]. Multiple passes of the laser beam can also be used to increase the absorption length; however care has to be taken that the laser crosses the beam almost perpendicularly, such that no additional Doppler broadening is caused. The sensitivity is limited by the bolometer noise. If the bolometer sees no stray laser light and is far from saturation, the signal-to-noise ratio will increase proportionally to the laser power. For this detection method, lasers with a small line width (≤ 1 MHz) but a reasonable power (≥ 1 mW) are used. The laser most often used in this type of experiment is the color center laser. A further development of this method by Fraser and Pine [52] included the use of focusing electric fields in the beam. Applications involve the use of buildup cavities in the beam for the observation of overtone transitions [56]. A different type of laser source for optothermal detection, the CO sideband laser, will be introduced in detail in Chap. 7.

5.4 Mass-Spectrometric Detection

If the molecular complexes in the beam dissociate upon laser excitation, the flux of the molecular beam is reduced, owing to the recoil of the fragments. This decrease in flux can also be detected by mass-spectrometric methods [24, 25, 84] or, alternatively, mass spectrometry can be used to detect the photofragments directly [196]. Molecular-beam electric-field resonance methods were extended to the infrared region by DeLeon and Muenter [40]. A general overview of deflection and focusing methods for molecular beams is given by J. Reuss in [173].

For the weakly bound complexes considered here, the mass spectrometer does not provide as much selectivity as is normally typical of such methods, owing to the considerable fragmentation that occurs upon ionization [16]. In order to obtain infrared spectra of truly size-selected clusters, a different approach was developed by Buck and coworkers [17]. A review of this work is given in [18]. Since mass spectrometers give only a lower limit for the cluster size, scattering with helium atoms was used to obtain an upper limit. The combination allows a determination of the actual cluster size.

Mass-spectrometric detection requires predissociation of the cluster, and therefore a high laser power is desirable. Since the line width of the transitions increases owing to lifetime broadening, lower-resolution spectra such as those obtained with a line-tunable CO_2 laser can be sufficient to obtain the complete spectra. More recently, experiments have also been reported which involve the use of an injection-seeded OPO [23, 88, 89].

This completes the general introduction. In the next two chapters we shall focus on the two spectrometers which were set up for the infrared spectroscopy of intermolecular complexes and discuss them in detail.

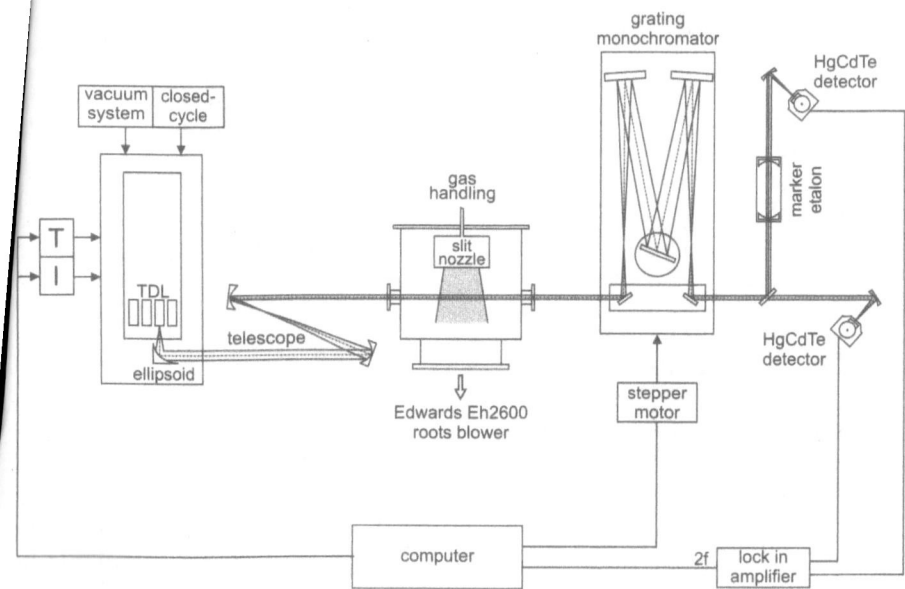

Fig. 6.1. Experimental setup of an IR diode laser spectrometer [99]

The frequency of the laser can be tuned by varying both the temperature and the current. A change of temperature will shift the gain profile of the laser and, at a constant current, will simultaneously change the index of refraction n within the laser diode. However, the tuning rate of the gain profile is a factor of 2–3 larger than the tuning rate due to the change of n. The combination of both effects leads to the following characteristic tuning behavior. The laser modes can be tuned continuously within certain limits (typically 1–2 cm^{-1}). At a certain temperature T the fast tuning of the gain profile will result in a mode jump, i.e. one mode terminates and a new mode at a shifted frequency starts to lase. An increase in the current I at fixed T has two consequences: First, n is changed, resulting in a negative tuning rate $\Delta\nu/\Delta I$. At the same time the temperature in the diode increases, even if the sink temperature is kept constant, which yields a positive tuning rate. In general, the second effect dominates, so that current tuning (T fixed) is similar to temperature tuning. Since current changes can be realized more quickly than temperature changes, tuning of the diode laser by current is normally used. Typical tuning rates are as follows:

- current tuning: $\frac{\Delta\nu}{\Delta I} = 0.2\text{--}3$ GHz/mA

- temperature tuning: $\frac{\Delta\nu}{\Delta T} = 10\text{--}100$ MHz/mK.

The laser diodes in our setup are cooled by a helium closed-cycle cooler (Leybold) down to temperatures of 20 K. The cryostat is mechanically iso-

6. The Diode Laser Spectrometer

6.1 Diode Lasers

Diode lasers are a technology whose importance is still increasing. Esp since the introduction of the "CD" player in the 1980s, they have bec great commercial interest. Owing to their efficiency and compactness have replaced gas lasers in some areas. While most commercial diode work in the red spectral region (ca. 635 nm), special diodes are availa the infrared and, more recently, in the green and blue regions (e.g. [5, In the following we shall focus on the use of an IR diode laser spectron for the investigation of weakly bound complexes.

A typical experimental setup for spectroscopy is shown in Fig. 6.1. B component of the spectrometer will be described in detail in the follow sections. A more general introduction to tunable-diode-laser spectroscopy given in [48, 109]. An introduction to the theory of lead salt diode lasers of be found in reference [97, 110].

6.2 The Laser

The IR laser system used in our experiments was a commercially available system (Mütek MDS 1100), consisting of laser diodes, a cryostat and optics. The laser diodes are lead salt diodes (from Laser Component and Aero Laser), where the active medium and the resonator are formed by the crystal. When an electric current is passed through a semiconductor diode, the electrons and holes recombine within the p–n junction and emit the recombination energy in form of infrared radiation. Lead salt diode lasers are available in the frequency range from 650 to 3300 cm^{-1}. One single diode can cover regions of 50–100 cm^{-1}. Lead salt diode lasers emit radiation with a typical line width of 30–100 MHz. Typical output powers are 100 µW to 1 mW. The single-mode tuning is limited to small regions, since the inhomogeneous broadening of the gain profile leads to multimode lasing. The actual quality of the laser diode depends on the individual diode (even diodes from the same crystal may vary in their characteristics), and on the specific frequency region. The laser noise shows a 1/f characteristic up to frequencies of 100 MHz.

lated against vibrations from the helium closed-cycle cooler. Up to four different laser diodes can be mounted on the cryostat and simultaneously cooled down. Temperature tuning is achieved via resistive heating. The heating element (a coil) touches the cold finger of the cryostat (a copper plate), which holds the diodes. A platinum resistor (Pt 1000) is used as a temperature sensor, which guarantees high stability, high absolute accuracy ($\pm$ 0.5 K), excellent reproducibility (0.05–0.1 K) and quick response. The combination of high-precision analog electronics for signal generation with sophisticated digital control techniques yields a temperature stability of 1 mK over time intervals of several hours. The current can be set between 0 and 900 mA via two 16-bit digital-to-analog converters, yielding steps as small as 0.3 µA. The current can be modulated using an internal source. The modulation frequency is typically 2–8 KHz, with a maximum amplitude of 1 mA. The amplitude is typically chosen such that the full modulation is less than the line width of the complex (ca. 50–100 MHz).

Diode lasers are very sensitive to any optical feedback, so the laser optics here consist exclusively of mirrors. The divergent laser beam is collimated by means of laser optics to a parallel beam of 14 mm diameter. The collimation optics consist of an ellipsoid with foci at 40 mm and 140 mm. The ellipsoidal mirror is adjusted in such a way that the diode is at the first focus. The laser radiation is transformed into a parallel beam of 14 mm diameter by means of a toroid ($f = 110$ mm). We use a telescope arrangement (two mirrors with $f = 60$ mm and $f = 10$ mm) to generate a parallel beam of only 3 mm diameter, which probes exclusively the expansion zone. In order to minimize astigmatic aberration, we have chosen nearly perpendicular incidence. The laser beam is directed into the expansion chamber via two adjustable mirrors.

6.3 The Vacuum Chamber

The vacuum is generated by a Roots blower (Edwards 2600 m^3/h), which is backed by a second Roots blower (Leybold 500 m^3/h) in combination with a mechanical pump (Leybold 65 m^3/h). The slit nozzle is 5 cm long and typically 50–70 µm wide. The backing pressure is typically 100–130 kPa, yielding a chamber pressure of 40–60 Pa. In order to increase the absorption path length the laser is multipassed within the vacuum chamber [100]. We have chosen a Herriot-type multipass arrangement [80, 81], in which each path crosses the narrow expansion zone. The laser beam is coupled into the cell by the edges of spherical mirrors placed in a near-concentric alignment. Using 50 mm diameter gold-coated mirrors (f = 75 mm), the optimum signal-to-noise ratio was achieved when the cell was aligned for 10–16 passes. This allowed an increase in the signal-to-noise ratio compared with the previous simple four-pass arrangement by a factor of 4–5. The laser leaves the vacuum chamber through a slightly tilted CaF_2 window and is then directed to the diagnostic and detection unit.

6.4 The Diagnostic and Detection Unit

Upon leaving the vacuum chamber, the laser is directed into a monochromator, used for mode separation and for a rough frequency determination (with an accuracy of about 1 cm^{-1}). The monochromator is a 50 cm grating monochromator (Mütek MDS 1200), using a 30 line/mm grating with a blaze wavelength of 25 μm (Carl Zeiss). The grating can be used over the entire spectral range of current interest. Absolute wavelength calibration is achieved using a He–Ne laser. While scanning, the monochromator is moved synchronously with the change of infrared frequency. Since current tuning rates can differ (from diode to diode) by one order of magnitude, the actual tuning rate has to be determined simultaneously. This is done by monitoring the transmission through the marker etalon. The computer control program uses this information to determine the tuning rate with an accuracy of better than 1% and readjusts the grating accordingly. A beam splitter (ZnSe window) directs some part of the laser beam into the marker etalon. The etalon is a stable confocal etalon, which generates a periodic frequency marker, serving to calibrate the spectrum. If required, a second beam splitter can be inserted which couples some part of the laser radiation into a reference cell, which serves for absolute frequency calibration. The various beams are focussed by gold-coated toroidal mirrors onto photoconductors (HgCdTe or InSb detectors). The laser noise exceeds by far (orders of magnitude) the detector noise, which implies that the laser noise limits the sensitivity of the detection. Therefore, 20–30% of the laser power is used for frequency calibration, since the signal-to-noise ratio is not decreased by the loss of laser power incident on the spectroscopy detector.

6.5 Mode Charts

Lead salt diode lasers can lase in several lateral modes. The gain profile is inhomogeneously broadened, typically allowing simultaneous operation in 3–5 modes. The homogeneous broadening can be comparable to the spacing between the modes, making mode competition possible. The presence of lateral modes and mode competition makes the mode structure complicated and unpredictable. Nevertheless, it turns out that this mode structure is rather stable and changes only slowly with time. In our group, M. Petri has developed a routine which automatically determines the mode structure, stores it and provides several possibilities to display this later, making the full information about the tuning range available for spectroscopy [153]. A typical mode chart is recorded on an equally spaced grid of roughly 100 distinct temperature and 100 current values (i.e. 10 000 operating points). The radiation frequency at each point is determined via the monochromator. The monochromator is scanned over the laser gain profile while, simultaneously, the transmitted intensity is recorded. These data are sorted into individual modes. Different temperature and current values belonging to the same mode are identified using estimated current/temperature tuning rates, which are iteratively improved. This procedure yields the complete mode structure and tuning rate for each laser diode. A mode chart of a typical diode is displayed in Fig. 6.2.

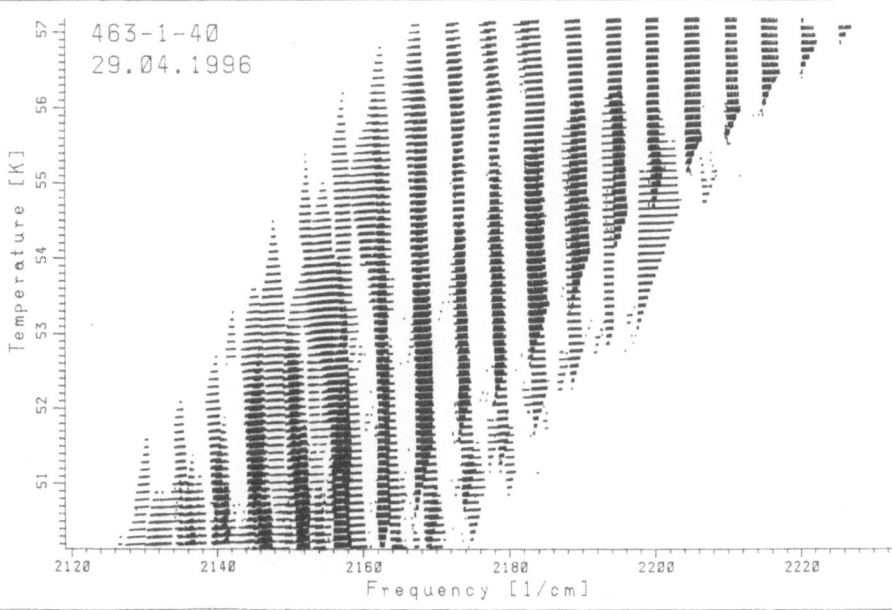

Fig. 6.2. Mode chart of a typical laser diode

6.6 Modulation Techniques

For spectroscopy, the signal is amplified and detected by phase-sensitive detection. Normally, frequency modulation (frequency ca. 8 kHz, amplitude ca. 10–50 MHz) is used, which is detected at $2f$. One inherent problem of laser diodes in connection with frequency modulation is that any small back reflection causes a periodic modulation of the baseline. This is enhanced by the inherent mode competition described in the previous section. An etalon structure appears in the spectrum, generated by the small intensity modulations. Since the detection limit in our setup is $\Delta I/I = 10^{-5} - 10^{-6}$ (with I being the intensity of the laser beam), reflections of the order of 10^{-5} will give rise to observable baseline modulations, which can hardly be distinguished from real lines. Even when special care is taken to avoid any back reflection, this problem is still present. The problem is enhanced if the back reflection leads to optical feedback in the laser diode. For this reason the monochromator was put as far from the laser diode as possible, since diffuse back reflections from the limited entrance slit could be effectively reduced by increasing the distance between the diode and monochromator to several meters.

An increase of the etalon structures was also observed in the case of multimode lasing. This indicated that optical feedback was still a cause of a problem.

7. The CO Sideband Spectrometer

7.1 The Experimental Setup

A chemically important frequency region is the region of the carbonylic stretch (1600–2100 cm^{-1}). So far, spectroscopic studies have been restricted to the resolution of lead salt diode lasers (50–100 MHz). However, the use of nonlinear frequency mixing of a finely tunable CO laser with microwave radiation allows the generation of narrow (< 300 kHz) IR radiation which can be used for high-resolution spectroscopy of complexes. The overall experimental setup of a such a spectrometer is shown in Fig. 7.1. The spectrometer consists of several components, which will be described individually below.

7.2 The CO Laser

The CO laser is a gas discharge laser with a high efficiency in the infrared region. The first laser activity was observed in 1965 by Patel [150]. The laser emits radiation from numerous vibrational–rotational transitions of the CO molecule. The pumping of populations into higher vibrational states is achieved via electronic excitation and the so-called "anharmonic vibration–vibration pumping". This process was studied for the first time by Treanor and Rich [189]. A partial population inversion is found for P transitions between adjacent vibrational levels, and laser emission can be achieved in the region from 1200 to 2100 cm^{-1}. The laser line width is less than 300 kHz. It is also possible to stabilize the CO laser using optogalvanic lamp dips [172]. An overview of recent results in CO laser development and its spectroscopic applications has been given by Urban [190, 191]. An extension of the available frequency region of the CO laser was accomplished by the development of the CO overtone laser by Urban and coworkers [64]. Laser emission could then be achieved for overtone (i.e. $\Delta v = 2$) transitions in the region between 2500 and 3800 cm^{-1} (for details see [4]). The spacing between nearby laser transitions is limited by the rotational spacing between adjacent J levels in CO, which amounts to ca. 4 cm^{-1}. For the CO fundamental laser we find an overlap between the different vibrational bands, since the spacing between adjacent vibrational bands is typically about 25–20 cm^{-1}, with decreasing

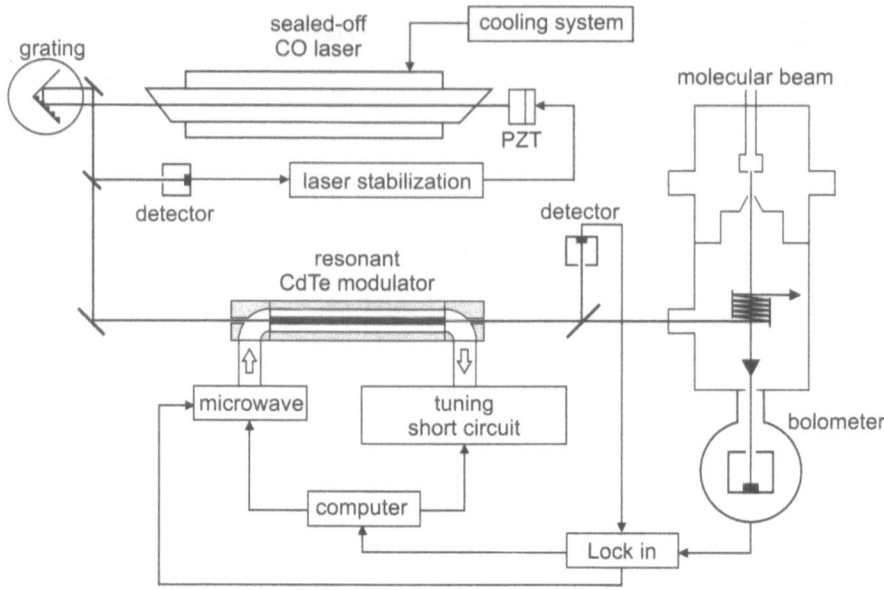

Fig. 7.1. Experimental setup of our CO sideband spectrometer using bolometric detection [130]

values for increasing vibrational quantum numbers. In order to increase the number of available laser lines, CO isotopes ($^{13}C^{16}O$ or $^{12}C^{18}O$) can be used. However, this requires the use of a sealed laser since a flowing laser system requires far too much gas for high-priced isotopes. The first operation of a sealed CO laser was reported in 1971 by Freed [54]. The inherent difficulty in the construction of sealed lasers is the cooling of the laser gas. In order to reach an effective population inversion, low plasma temperatures are required. For flow lasers the discharge tube is cooled by liquid nitrogen (80 K). Laser operation can be achieved at room temperature [10], but only for a limited number of lines and a smaller laser output power. However, the use of liquid-nitrogen cooling in sealed systems results in termination of the laser emission after a short period. This can be explained by chemical reactions which decrease the percentage of CO in the gas mixture. For a constant laser output power over longer periods, higher plasma temperatures (≥ 110 K) are required. However, in order to achieve optimum conditions for laser action, temperatures of less than 170 K are desirable.

To meet these conditions, a new design to provide variable plasma temperatures was developed [129, 139]. By inserting an additional layer between the discharge tube and the liquid-nitrogen reservoir, a temperature gradient between the plasma and the liquid nitrogen is established. The additional layer should have a variable thermal conductivity. This is achieved by a sand–air mixture, which works as follows. The thermal conductivity for an ideal gas is proportional to the density n and to the mean free path l. For a gas, n is proportional to the pressure P, whereas l is inversely proportional to P. This implies that the thermal conductivity of air, for example, is independent of P. However, if we add sand particles and reduce the pressure we can reach a point at which the mean free path is limited by the mean distance between the sand particles. Accordingly, the mean free path will be independent of P. This has the consequence that the thermal conductivity of a sand–air mixture is directly proportional to P. We can therefore adjust the thermal conductivity by variation of P, which allows control of the plasma temperature over a wide temperature range (100–300 K).

As the discharge tube, an Al_2O_3 ceramic tube, with an inner diameter of (11.1 ± 0.3) mm was chosen. The actual diameter was a compromise between the requirement of a small diameter for a low central plasma temperature and of a large diameter in order to prevent a perturbation of the laser beam, given a beam waist of 2–5 mm. The temperature for optimum laser performance, as measured outside the discharge tube, is 100–150 K, depending on the laser line. The optimum gas mixture is 3% xenon, 6% carbon monoxide, 10% nitrogen and 81% helium at a total pressure of 20 hPa. For laser transitions involving low vibrations, a smaller CO percentage should be chosen (see [57] for the performance of this laser in the fundamental band). The laser resonator consists of a Zeiss grating (250 lines/mm) in the Littrow configuration and a 5 m end mirror. In order to establish a stable discharge, a resistor (here, 400 kΩ) is inserted between the high-voltage supply and the discharge. Typical discharge conditions are $I = 10$ mA and $U = 15$ kV. The maximum (multimode) output power is 7 W. Using three isotopes, laser operation can be achieved for ca. 400 lines in the frequency region 1566–2037 cm^{-1}. This corresponds to an average spacing between two laser lines of 1.2 cm^{-1}. The tuning of the CO laser lines is restricted to the gain profile of the laser, which corresponds to 120 MHz. The spectral coverage in the infrared region is therefore less than 1%.

7.3 The CO Sideband Laser

A CO sideband laser requires a nonlinear element in which CO laser radiation and tunable microwaves are mixed and sidebands are generated. The sidebands provide radiation at the sum and difference frequencies and allow tuning of the infrared radiation. The technology of sideband generation using a powerful IR laser and an electro-optical modulator was first applied to spectroscopy by Cocoran et al., who measured CO_2 gain profiles by scanning them point by point with tunable sideband generation [34]. Similar point-to-point spectra were obtained later by Magerl and coworkers [121]. The first sub-Doppler spectra obtained using sidebands in an optical double-resonance scheme were reported by Oka and coworkers [148]. Magerl et al. improved the sideband efficiency by nearly three orders of magnitude by solving the phase-matching problem, which extended this technique to continuously tunable sideband generation in the CO_2 laser region [122, 123, 124]. They achieved tunable sideband generation over regions 23 GHz wide above and below each of some 80 CO_2 laser lines. This technique was transferred to the CO laser by Hsu et al. [86]. Further applications were reported by the Laboratoire de Spectroscopie Hertziénne in Lille [31] and by a cooperation between the groups in Lille and in Bonn [131].

The principle is as follows. The laser is modulated using the electro-optical effect in a CdTe crystal. The otherwise isotropic crystal becomes birefringent when an external electric field is applied. For modulation in the GHz region, microwaves are used to provide the electric field. The linearly polarized laser beam is then split into two components which propagate with different velocities in the crystal. After leaving the crystal, the two components show a phase difference, which is time-dependent owing to the time dependence of the microwave field and corresponds to a rotation of the polarization. In a first-order approximation we obtain two sidebands, which are both polarized perpendicular to the polarization of the incoming CO laser beam, the carrier. The carrier is effectively suppressed by the use of a polarizer, which is oriented perpendicular to the initial polarization but leaves the sidebands unchanged.

The efficiency η of the sideband generation is given by

$$\eta = \left(\frac{\pi n_0^3 r_{41} L}{2\lambda_{\text{laser}}} \right)^2 E_{\text{MW}}^2 \frac{\sin^2 x}{x^2} , \tag{7.1}$$

where n_0 is the refractive index, r_{41} the component of the electro-optical tensor which describes the change of n due to the influence of an electric field, L the length of the crystal, λ_{laser} the wavelength of the laser and E_{MW} the electric field of the microwave radiation. The parameter x is given by

$$x = \frac{\omega_{\text{MW}} L(1/u_{\text{laser}} - 1/u_{\text{MW}})}{2} , \tag{7.2}$$

where ω_{MW} is the microwave frequency and $(1/u_{\text{laser}} - 1/u_{\text{MW}})$ describes the velocity match between the microwave and laser propagation velocities in the crystal. The sideband power is proportional to η and to the incident CO laser power. We can see from the above expression that the sideband power is proportional to the microwave power and to the length L. It is inversely proportional to the wavelength of the laser, which explains why the efficiencies reached in the IR region are much smaller than the efficiencies usually reached in the optical region. This dependence also explains why the conversion to sidebands is more effective for the CO laser than for the CO_2 laser. We can see, further, that effective sideband generation requires a velocity match of the laser beam and microwave propagation in the crystal. Only when such a velocity match is achieved (i.e. $u_{\text{laser}} = u_{\text{MW}}$) will sidebands generated in different parts of the crystal interfere constructively to give an increase in the sideband power proportional to L. However, this restricts the maximum length of the crystal (in our setup, 40 mm). One design used by many groups is the so-called Magerl design (see [86] and references therein). Velocity matching of the laser and microwave radiation is achieved by choosing the right dimensions for the microwave waveguide. The rectangular waveguide in this design is optimized for 13 GHz and can be used from 8 to 18 GHz. The CdTe crystal is 3×3 mm in size and has an antireflection coating. In order to reach the optimum waveguide dimensions of 3×6.8 mm with a uniform dielectric constant ϵ, the space is filled with Al_2O_3, which has nearly the same ϵ as CdTe (9.9 instead of 10) but is much cheaper. The microwaves are generated by a microwave synthesizer and amplified to ca. 20 W. The microwaves can be chopped by a p–i–n diode, which is inserted between the synthesizer and the traveling-wave-tube amplifier (TWTA). The microwaves are coupled into the modulator by a double-ridge waveguide, as indicated in Fig. 7.1. The double-ridge waveguide allows efficient microwave transmission in the region 8–18 GHz. The bent double-ridge waveguides used here have small circular holes (diameter 4 mm) to couple the IR laser beam into the crystal. The microwaves travel through the waveguide, which contains the crystal and is then terminated by a power load.

Typical operation conditions for such a setup are shown in Fig. 7.2. The maximum output power of each sideband is 0.25 mW per 0.5 W incident laser power. This maximum power is reached at 13 GHz and decreases to ca. 0.02 mW at 8 GHz and 0.13 mW at 18 GHz (all values per 0.5 W incident laser power). On top of this broad structure, which roughly follows theoretical predictions taking account of velocity matching, we see sharper resonances. These substructures can be explained by imperfect matching between the different microwave components, e.g. the double-ridge–rectangular waveguide connections, which lead to microwave back reflections. The available sideband power can be further reduced by imperfect overlap between the microwave and IR laser radiation or by losses due to air gaps between the crystal and the surrounding waveguide.

A different modulator design was developed by Cheo [30]. Instead of waveguides, his design involves integrated optics. He uses a CdTe-buffered GaAs thin-slab waveguide modulator for sideband generation. The output power varies dramatically for different modulators, owing to the complicated technique. One disadvantage of this technique is that the carrier and sideband have the same polarization and cannot be separated easily. In the case of absorption spectroscopy, this implies that the noise of the carrier will contribute to the noise at the detector and will exceed the noise of the sidebands by far, since the carrier typically has 1000 times more power. Using the Magerl design, the carrier and sideband can be easily separated by a polarizer. After the polarizer, we reach typically a sideband–carrier ratio of 1:1. In the case of optothermal detection the carrier is not such a severe problem,

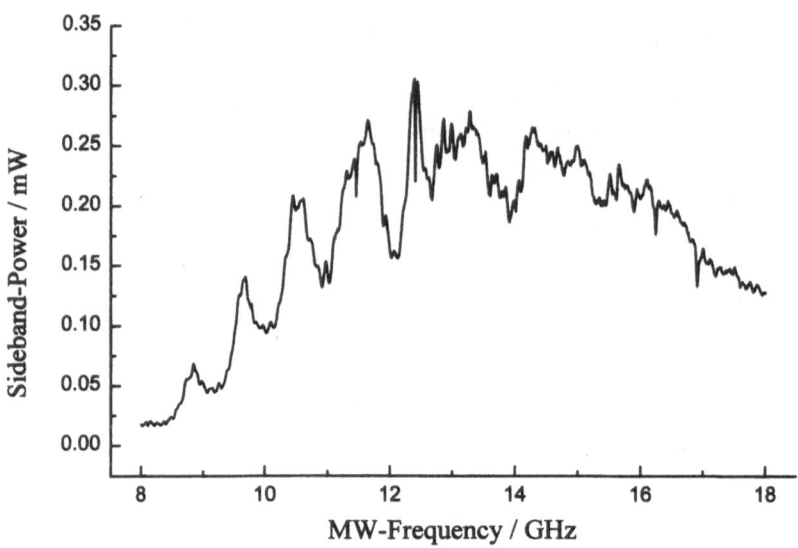

Fig. 7.2. Sideband power of each sideband per 0.5 W laser power

since the molecular beam and not the laser light is detected. However, stray light from the IR laser source and the bolometer detector might become a serious problem, especially in the case of a high-power IR laser. Nevertheless, the first application of sideband lasers to optothermal detection was reported by Fraser et al. [53], who observed the infrared spectrum of Ar–NH$_3$ using a CO$_2$ and N$_2$O sideband laser and a modulator of the type developed by Cheo. Using 20 W of microwave radiation and 2.5–6 W of IR radiation, 1–2 mW of tunable infrared radiation could be generated.

In general, the output powers of 0.25 mW expected for CO sideband laser spectroscopy are not sufficient for bolometric detection of molecular complexes. Instead, an output power of 5–10 mW, which corresponds to a typical saturation power, would be ideal, and an output power of 1–2 mW would be considered a minimum for wide application of this technique. We therefore had to improve the output power, which could be achieved by the construction of a resonant CO sideband laser.

7.4 The Resonant CO Sideband Laser

According to (7.1), the sideband power is proportional to the CO laser power and the microwave power. Hence, an increase of the sideband power can be achieved by the use of optical or microwave resonators. An optical resonator suffers relatively high losses due to the crystal. Typical losses resulting from the coupling of the CO laser beam through the small (3 × 3 mm) crystal, when using a broadband antireflection coating on each side of the crystal, are on the order of 20% per round trip. For a perfect optical cavity the maximum gain would therefore be on the order of 4–5. However, the most effective and simple approach is to build a resonator for the microwave radiation. Microwave cavities have the advantage that they require much less sophisticated tuning. As we shall see later, no additional stabilization is necessary.

The first design of a (microwave) resonant modulator as an alternative to the traveling-wave modulator was described in [124]. The power load was replaced by a tunable short. Coupling into the resonator was accomplished via cutoff coupling [14]. For cutoff coupling, a tapered, ridged guide is inserted in front of the modulator. This has the same dimensions as the traveling-mode resonator; however, it lacks any dielectric medium inside ($\epsilon = \epsilon_0$), and thus causes an exponential damping of the traveling microwave. The coupling efficiency can therefore be adjusted by means of the length of the guide. However, for efficient coupling, a readjustment of the alumina slabs inside the modulator was required. This setup gave rise to several microwave resonances spaced 250 MHz apart. As a consequence, the modulating electric field in the modulator was enhanced by the microwave resonator. However, if we compare the sideband power for these two setups, one has to keep in mind that only the microwave power in the modulator will contribute to the sideband generation. Furthermore, we have a standing wave in the resonator, which implies that only one half (the half which travels in the same direction as the laser beam) can be used, since only this part is velocity-matched. For the traveling wave, the square of the modulating electric-field strength is given by

$$E^2 = \frac{4P_{MW}Z_{TE10}}{ab} , \qquad (7.3)$$

where P_{MW} is the power of the microwave radiation Z_{TE10} is the impedance of the modulator and a, b describe the dimensions of the modulator waveguide. In comparison, the square of the effectively usable modulating electric field for the resonant modulator is given by

$$(E_M)^2_{eff} = \frac{2QP_{MW}}{\omega\epsilon_0 ab\left(\epsilon_r L + L_T Z_T/Z_M\right)} , \qquad (7.4)$$

where Q describes the unloaded Q factor of the tuner–modulator combination, ω_M is the microwave frequency, ϵ_r is the dielectric constant of the modulator, L is the length of the modulator, and Z_T and Z_M are the impedances of the resonator and tuner, respectively [123]. The maximum output reported was 1 mW sideband power for each watt of incident laser power.

For spectroscopic applications, the resonator should be automatically adjustable over the frequency range of 8–18 GHz. A schematic setup of a scanning CO sideband laser using resonant microwave radiation is shown in Fig. 7.1. The adjustment is realized by a sliding short, which can be continuously reset by a micrometer screw, driven by a stepping motor. The tunable short consists of a copper plate which moves inside a double-ridge waveguide. The plate can be tuned over 50 mm inside the waveguide with an accuracy of 1 μm. The maximum required tuning range corresponds to a change of 30 mm in the resonator. The end of the waveguide is filled with a microwave absorber in order to prevent back reflection from the fixed end. The coupling of the microwave into the cavity is accomplished via a copper plate with a pinhole of 7 mm diameter. This plate is mounted in front of the bent double-ridge waveguide; this arrangement turned out to give the best performance. Other designs, such as different pinhole diameters, other shapes such as vertical and horizontal slits, and changes in the adapter between the double-ridge and the microwave cable were tested, but did not result in any improvements.

If the microwave cavity is adjusted for the selected microwave frequency, the output power of the sidebands is increased by a factor of 2.5 to 5, depending on the microwave frequency. The tuning of the microwave cavity can be achieved either by optimizing the CO sideband laser power or by minimizing the microwave reflections. The microwave reflections are measured by a crystal detector behind a minus 30 dB directional coupler, which is inserted between the TWTA and the microwave resonator.

For continuous tuning of the sideband laser, the resonator is mechanically stepped forward until the microwave resonances are minimized. In practice, the resonator is tuned in fine steps over the resonance curve and reset if more than 40 mm tuning is required. The optimized position of the short circuit is recorded as a function of the microwave frequency. The position of the minima turns out to be very reproducible. It is therefore sufficient to record the position of these minima once, before the CO laser is actually started. During normal operation, the microwave cavity is tuned according to these values via computer control without any further adjustment. A set of results is shown in Fig. 7.3 [49].

In the first step, the microwave power was adjusted for the optimum performance of the TWT amplifier. The lower trace of Fig. 7.3 shows the resulting sideband power at the optimized microwave power. In the upper trace of Fig. 7.3, the result obtained using the resonant modulator is shown. The substructure in the gain profile is not a result of imperfect tuning of the resonator, but solely due to the imperfect coupling of the microwaves into the crystal and is caused by reflections at various connections. Between 10 and 13 GHz, optimum output powers of 1.8 mW sideband power per 0.5 W incident CO laser power are obtained.

We can compare this result with the result of the Magerl group. This group obtained a maximum of 1 mW per 1 W IR laser power if the tuner was set at a fixed position. However, if we take into consideration the fact that the efficiency η increases for the CO laser as $1/\lambda_{laser}^2$, we can conclude that the efficiencies are comparable. On the other hand, continuous scanning is possible in our setup. We can also compare our setup with the results of

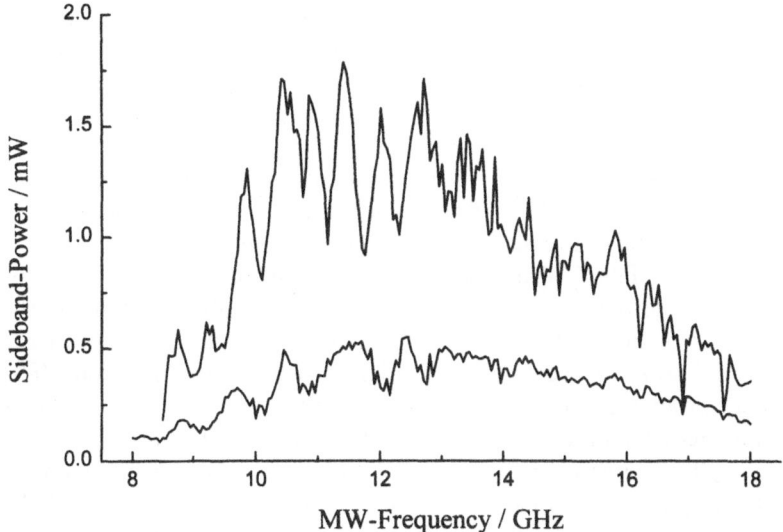

Fig. 7.3. Sideband power of each sideband per 0.5 W laser power for the microwave resonant modulator

the Cheo design, which was applied to spectroscopy by Fraser et al. [53]. In the CO_2 laser region, these authors obtained an output power of 1–2 mW in the sideband with 2.5–6 W IR radiation. In conclusion, the resonant Magerl design yields a higher output power than the Cheo design.

The improvement compared with our previous result (7.2) is at least a factor of 6, but can be as high as a factor of 10. The resonator gives an improvement of a factor of 2–3 on the average and up to a factor of 5 for certain frequencies. If we assume a typical dielectric Q value of 600–700 for CdTe, neglect the wall losses in the double ridge and assume an ideal coupling of the microwaves in the resonator, we expect from a comparison of (7.3) and (7.4) an increase of a factor of 5–6. We can see that we are close to ideal conditions.

For a typical CO laser output power of 1 W, we obtain 2–3 mW sideband power for each sideband. The spectral coverage is approximately 50% (including the use of the various CO isotopes). The line width is less than 1 MHz, which is a precondition for the spectroscopy of molecules in a skimmed jet. This enables, for the first time, the application of bolometric detection in the chemically important region of the carbonyl vibration between 1600 and 2000 cm^{-1}.

7.5 The Vacuum Chamber

The differentially pumped vacuum system in our setup consists of three chambers. The first chamber is pumped by a 6000 l/sec diffusion pump. This is backed by a 250 m^3/h Roots pump, backed by a 46 m^3/h mechanical pump. The first chamber contains the point nozzle. The diameter of the nozzle can be varied; typically, diameters of 50 to 100 μm are chosen. With an inlet pressure of 100 kPa and using a 1–3% mixture of the gas in helium, the chamber pressure is in the region of 1×10^{-2} Pa.

The molecular beam enters the second chamber via a skimmer with a diameter of 0.5–1 mm, depending on the nozzle diameter. The distance from the point nozzle to the skimmer is 25 mm. The chamber is pumped by a 450 l/s turbo pump. The typical background pressure is 4×10^{-4} Pa. This chamber includes the optics for the laser beam. The multipass cell for the laser consists of two gold-coated mirrors of 30 mm diameter. The two mirrors are fixed parallel to each other and are aligned parallel to the beam. The sideband laser is focused before entering the chamber by the use of a mirror ($f = 1$ m). The laser beam has a diameter of 2 mm when entering and leaving the multipass cell; the beam waist inside is 1 mm. When the laser is coupled into the cell, the beam enters the cell as close as possible to one side of the mirror in order to make the angle at which the laser beam and the molecular beam cross almost 90°. The actual deviation amounts to 3°.

The third chamber consists of the cryostat containing the bolometer detector (Infrared Laboratories). The dewar can be filled with 2.8 l of liquid helium and 2.5 l of liquid nitrogen. This chamber is pumped by a 350 l/s turbo pump. Both turbo pumps are backed by a mechanical 30 m^3/h pump. If the cryostat is filled, the cryogenic pumping exceeds the pumping of the turbo pumps. The typical pressure in this chamber is 1×10^{-5} Pa. The distance from the nozzle to the bolometer detector is ca. 550 mm. The detector element has a surface area of 2 mm × 5 mm. For optimum performance the bolometer has to be cooled down to 1.6 K. This is accomplished by pumping the helium with a mechanical pump below the λ point. The performance of the bolometer will be described in more detail in Sect. 1.1.

7.6 The Expected Line Width

The expected line width is limited by several contributions. The time-of-flight broadening can be calculated according to

$$\delta\nu_t = \frac{\sqrt{2\ln 2}}{\pi} \frac{v}{\omega_0},$$

where v is the velocity of the molecular beam (i.e 1765 m/s for helium) and ω_0 is the laser beam waist (i.e. 1 mm). The time-of-flight broadening can therefore be estimated as $\delta\nu_t = 0.7$ MHz.

The Doppler broadening, in the case of perpendicular crossing of the laser beam, is limited by the collimation of the beam. The factor which restricts the possible velocity component perpendicular to the beam ($v_\perp$) is the bolometer itself (and not the skimmer). The maximum possible $v_\perp$ which will reach the detector is given by $v_\perp = v(1/2)(5\ mm/550\ mm)$. The corresponding Doppler width in our setup is $\delta\nu_\perp = 1.6$ MHz.

Owing to the multipass arrangement, perfect perpendicular crossing of the laser beam with the molecular beam is not possible. The deviation of $\theta = 3°$ causes an additional Doppler broadening. For a description of this broadening we have to distinguish between $T_\parallel$ and $T_\perp$, which describe the velocity distributions in the direction of the beam and perpendicular to it, respectively. In general, $T_\parallel$ exceeds $T_\perp$ (see the chapter by R. Miller in [173]) since $T_\parallel$ remains constant or frozen in the collisionless regime. The spread of velocities described by $T_\parallel$ for a beam with a temperature $T_\parallel$ of 5 K is $\delta v_\parallel = 145$ m/s. If we cross the beam at an angle θ, we have a Doppler broadening of

$$\delta\nu_{\mathrm{Dopp}} = \delta\nu_\parallel \sin\theta + \delta\nu_\perp \cos\theta .$$

If we neglect the second contribution, i.e. that due to $\nu_\perp$, which is small in comparison with the first part, we obtain $\delta\nu_{\mathrm{Dopp}} = 1.4$ MHz. Taking into account all contributions, a broadening of $\delta\nu = 2.2$ MHz is expected, which agrees well with the experimental value of 2.5–3 MHz.

7.7 The Detection Unit

A detailed review of cryogenic bolometers was given by M. Zen in [173]. Here we shall focus on the performance and limitations of a specific setup and discuss the limitations. Bolometers are commercially available (Infrared Laboratories). The typical sensitivity is $S = 9.5 \times 10^5$ V/W, with a noise equivalent power (NEP) at 80 Hz of 4.3×10^{-14} W/$\sqrt{\mathrm{Hz}}$. The NEP describes the energy of the beam required to generate a signal which is comparable to the inherent noise of the bolometer. For an integration time of 1 s this NEP corresponds to 9×10^5 excited molecules (at 2000 cm^{-1}) or, taking into account the thermal conductivity of the crystal ($G = 8.6$ µW/K), a thermal variation of $\delta T = 10^{-8}$ K. The flux of the molecular beam at the centerline per steradian is given by

$$F = n_0 c\kappa \frac{\pi d^2}{4} \sqrt{\frac{2kT}{m}} ,$$

where c, is a constant which depends on γ ($c = 0.5$ for monatomic gases), κ is the peaking factor ($\kappa \approx 2$ for monatomic gases), and n_0, T and m are the density, temperature and mass of the gas in the source, respectively [173]. For typical beam conditions of $p_0 = 100$ kPa and $d = 50$ µm, we have a flux of 2×10^{19} molecules/(s steradian) for a helium beam. At a bolometer ($x = 550$ mm) with an area of $A = 10$ mm^2 we therefore have a flux of

$$f = F_n \frac{A}{\pi x^2} = 5 \times 10^{14} \text{ molecules/s} .$$

A typical molecular beam provides a flux of approximately 10^{14} molecules/s. Whereas lower fluxes yield smaller signals, a higher flux no longer leads to an increase in signal-to-noise ratio. For higher fluxes, the shot noise exceeds the bolometer noise (increasing linearly with the flux) and thereby limits the sensitivity of the detection, whereas for lower fluxes, the inherent noise sources of the bolometer exceed the shot noise of the beam. Moreover, the molecular beam heats up the bolometer. This results in a decrease of the bolometer sensitivity for a large molecular flux.

For an estimate of the maximum expected signal-to-noise ratio we have to take into account the following numbers. In order to obtain the number of molecules which can actually be excited in the beam, we have to know the composition of the mixture, i.e. the ratio of spectroscopic gas to carrier gas (r), and the percentage of molecules in the state i (r_i) which is actually excited. This is given by

$$r_i = \frac{g_i \exp\left(-E_i/kT\right)}{\sum_{j=0}^{\infty} g_j \exp\left(-E_j/kT\right)} ,$$

where g_j is the degeneracy, E_j is the energy of the state j and T is the temperature. The maximum signal-to-noise ratio is achieved if 50% of the molecules in this particular state are excited by the infrared radiation, which corresponds to complete saturation (reducing r by a factor of 2). Typically 5–10% of a strong IR absorber (e.g. HF) is excited. The maximum signal-to-noise ratio is given by

$$\frac{S}{N}_{\text{max}} = \frac{frr_i}{9 \times 10^5 \text{ molecules/s}} \frac{\sqrt{\tau}}{\sqrt{s}} ,$$

with τ being the integration time. As a test of the CO sideband spectrometer the P(1.5) transition of NO was measured using a mixture of 3% NO in helium. For optimum performance (5% excitation, $T = 3$ K, $\tau = 1$ s) we expect an S/N ratio on the order of 10^4. The measured signal-to-noise ratio of the strongest hyperfine component ($\tau = 100$ ms) is 1000, which is close to the expected value and demonstrates the high sensitivity of this instrument. This setup allows for the first time high-resolution spectroscopy in the carbonyl region. As an initial result the fully resolved IR spectrum of the formic acid dimer was obtained [130].

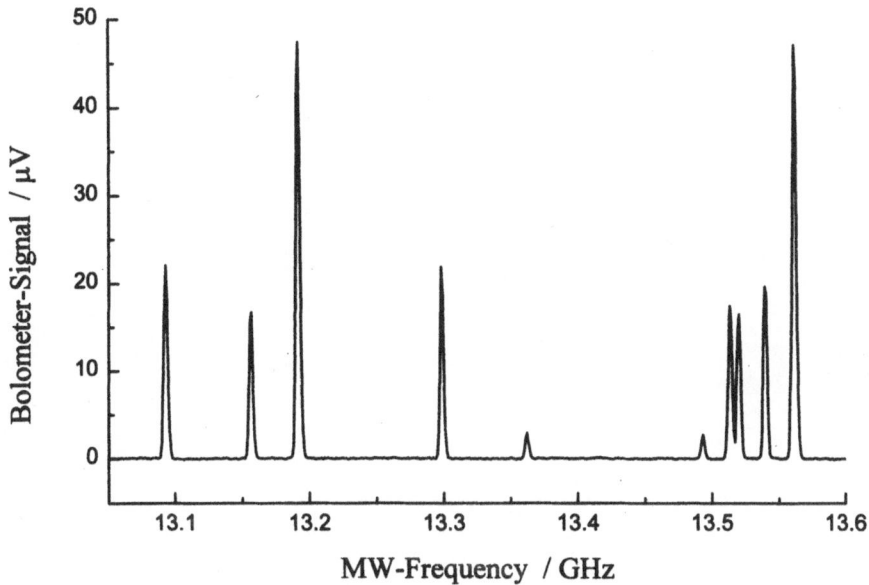

Fig. 7.4. Bolometric detection of the P(1.5) transition of NO. The λ doubling and the hyperfine structure are resolved

8. Infrared Spectroscopy of Ar–CO

8.1 Motivation

We shall now investigate one prototype system for intermolecular interaction, the Ar–CO system, in more detail. In Chapter 1 of this book I described the different contributions to the intermolecular potential. I furthermore discussed one example, the Ar–HF complex, which has been studied by several groups. Complexes consisting of a noble gas and a diatomic molecule are the simplest systems in which anisotropic intermolecular forces play an important role. They are especially suitable for testing recent theoretical developments since a direct comparison with ab initio studies is possible. For experimental reasons, nearly all detailed studies of small complexes involve hydrogen-containing molecules; the complexes studied include Ar–HCl, Ar–HBr, Ar–HF, Ar–NH$_3$0 and Ar–H$_2$O. The molecules have large dipole and transition dipole moments and are therefore among the easiest to detect. For all these complexes the induction energy makes a considerable contribution to the overall binding energy, especially to the anisotropic part of the potential. The equilibrium structures of the complexes Ar–HBr, Ar–HCl and Ar–HF show a linear structure, corresponding to the maximum in the induction energy. However, the induction energy is the part of the intermolecular interaction (and in addition to the electrostatic part) which can be most easily predicted. Reliable estimates can be obtained from the known polarizibilities and electrostatic moments (μ and Q) of the monomers [90]. An accurate description of the parts involving correlation effects, such as the dispersion interaction, is theoretically more challenging. The estimates are based on knowledge of the C_0^6 coefficients [200] and on the measured anisotropic polarizibilities of the molecules [44]. For Ar–CO the induction energy is nearly negligible (less than 2% of the overall binding energy). The potential and especially the anisotropic part of the potential surface are therefore determined exclusively by a balance of dispersion and repulsion. A detailed knowledge of the potential surface therefore allows access to these two contributions and a direct test of theoretical models for repulsion and dispersion, without any further corrections.

The only other nonhydride complex for which several van der Waals states have been measured is the He–CO complex, which has been studied by McKellar et al. [29, 33]. A pure ab initio prediction of the infrared spectrum, which coincides very well with the experimental spectrum, is given in [136]. However, for He–CO only a few bound states can be found in the flat potential ($D_e = 22.9$ cm^{-1}; the lowest energy level is bound by only 6.6 cm^{-1}). As a prototype system He–CO therefore lacks a useful number of possible experimental data values. For Ar–CO, binding energies of the order of 110 cm^{-1}, and hence a much higher number of bound states (50–100), are predicted. Compared with the hydride complexes, which have even higher binding energies, the Ar–CO complex has an increased number of bound states owing to the small rotational constant b_{CO} of CO (as compared with b_{HF} for Ar–HF). This rotational constant b determines the energy separation of different rotor states for the CO complex within the Ar–CO complex and is therefore responsible for the density of van der Waals states in the potential. For these reasons, Ar–CO can serve as an excellent prototype system for the study of van der Waals complexes and for the investigation of intermolecular interactions. Moreover, Ar–CO has turned out to be one of the best studied molecular complexes. More than 20 papers have already dealt with this prototype, which makes it one of the best studied molecular complexes lacking any H or D atom. In the next section I shall give a summary of the previous measurements of other groups, and present in later sections the results of our own infrared studies, which were performed using the diode laser spectrometer described in Chap. 7.

8.2 Previous Studies

The first observation of infrared spectra of Ar–CO was reported by Piante et al. [42]. However, the infrared spectra were not analyzed, but served only to demonstrate the sensitivity of their new infrared spectrometer. The first detailed spectroscopic study of Ar–CO was reported by McKellar et al. [126]. Using a long-path low-temperature static gas cell in combination with a Fourier transform spectrometer and a pulsed slit supersonic expansion in combination with a tunable diode laser, they were able to observe several infrared transitions of Ar–CO. The infrared photon excites the CO monomer vibration. The excitation energy is far above dissociation; however, owing to the extremely small coupling between the monomer vibration and the van der Waals modes, no predissociation is observed. The overall band origin was estimated to be at 2142.83 cm^{-1}, which corresponds to a fairly small shift of -0.44 cm^{-1} with respect to the unperturbed CO vibration. The molecular parameters were found to depend very little on the CO excitation, reflecting the small coupling between inter- and intermolecular modes. Although the temperatures in the cell (80 K) or in the jet (15 K) were fairly cold, several bound states in the intermolecular potential could still be populated and could serve as a ground state. In general, combination bands of CO vibration and van der Waals modes can be excited, yielding valuable information on the intermolecular potential for $v_{CO} = 0$ and $v_{CO} = 1$. McKellar et al. were able to assign six b-type perpendicular ($\Delta K = 1$) sub-bands in the CO fundamental branch of Ar–CO. These authors determined the energies of states which correspond to the out-of-plane rotation of the CO monomer (see Sect. 3.4). However, no van der Waals modes were reported. The rotational spacing of the sub-bands yielded the overall rotational constant B of the complex. Analyzing this in terms of a rigid molecule, the complex was found to be a T-shaped nearly prolate asymmetric rotor molecule with a B constant of 0.0691 cm^{-1} in the ground state. This implies an average intermolecular separation of 3.85 Å.

Measurements of pure rotational spectra of various isotopomers of Ar–CO in the $v_{CO} = 0$ ground state were published by Ogata et al. [145]. These measurements allowed a precise determination of the rotational constant in the ground state. Depending on the underlying model, Ogata et al. obtained values between $80.2°$ and $69.5°$ for the angle θ between the intramolecular and intermolecular axes (with the oxygen closer to the argon). These variations indicate that the use of a rigid-rotor-type model for Ar–CO might not be justified. On the basis of the distortion constant Δ_J, the energy of the stretching mode was estimated to be 27.2 cm^{-1}. More recently, the A rotational constant of Ar–CO was directly determined by the observation of b-type transitions [79, 95].

On the basis of a potential surface obtained by Mirsky ($D_e = 110$ cm^{-1}, $R_e = 3.63$ Å, $\theta_e = 77°$) [134], the first detailed theoretical calculations of the bound energy levels of Ar–CO were performed by Tennyson et al. [186]. These calculations predict the bending energy to be located at 14.3 cm^{-1} and the energy of the van der Waals stretch near 24.2 cm^{-1}. Giving special attention to the onset of irregular behavior in the spectra due to vibration-rotation interaction (Coriolis coupling), Tennyson et al. found regions near the dissociation limit and close to accidental degeneracies where the normal regular overall rotational pattern of Ar–CO is completely destroyed. Parish et al. [149] (using $D_e = 158$ cm^{-1}, $R_e = 3.81$Å and $\theta_e = 49°$) obtained an energy of 13 cm^{-1} for the bend and 42 cm^{-1} for the stretch using a harmonic approximation, which, however, is not valid for floppy van der Waals complexes.

On the basis of an unpublished potential obtained by Shin ($D_e = 69$ cm^{-1}, $R_e = 4.0$ Å, $\theta_e = 100°$), Castells et al. [26] predicted a bending energy of 7.8 cm^{-1} and a stretching energy of 27.6 cm^{-1}. The underlying potential used second-order Møller–Plesset perturbation theory (MP2); the binding energy was calculated to be much lower (69 cm^{-1}) than in the previous potential calculations by Mirsky (110 cm^{-1}) and Parish (158 cm^{-1}). The calculations of Castells et al. are based on a Coriolis decoupling approximation, coupling of nearby states are not taken into account. In addition to the calculation of the bound states, the intensities of the various infrared transitions were predicted. A similar result for the binding energy was derived by Jansen using supermolecular coupled-pair functional calculations ($D_e = 71$ cm^{-1}, $R_e = 3.9$ Å,

$\theta_e = 92°$); however, this method is known to underestimate the binding energy (for Ar_2, only 70% of the true well depth was obtained using this method). A further potential surface was published by Kukawska-Tarnawska et al. [103], which was calculated using MP4 methods ($D_e = 109$ cm^{-1}, $R_e = 3.75$ Å,$\theta_e = 80°$). However, no calculations of bound states based on this potential have been presented. An independent MP4 calculation was performed by Shin et al. [176] which yielded similar values for the equilibrium structure as described by R_e and θ_e; however D_e was found to be slightly decreased ($D_e = 96$ cm^{-1}, $R_e = 3.74$ Å, $\theta_e = 82°$)compared to the MP4 calculations by Kukawska-Tarnawska et al. [103].

In 2000 a potential energy surface has been developed by Toczylowski and Cybulski [187] using single and double excitation coupled-clusters theory with non-iterative treatment of triple excitations [CCSD(T)]. They obtained a remarkable agreement with the experimental data for intermolecular bending and stretching modes. Whereas the experimental values are 12.014 and 18.110 cm^{-1} theoretically values of 11.894 and 18.175 cm^{-1} were predicted using CCSD(T).

A similar good agreement was obtained by Gianturco and Paesani [58] who combined density functional theory (DFT) with long-range dispersion to obtain adiabatic surfaces. They obtained 11.649 cm^{-1} and 18.933 cm^{-1} for the bending and stretching mode in $v_{CO} = 1$. The global minimum was -98.6 cm^{-1} located at $R = 3.81$ Å and $\theta = 81°$.

In addition to these pure ab initio potential surfaces one adjusted potential surface has been published by Jansen [96] and will be discussed within the next chapter

8.3 The Bending Excitation in Ar–CO

In Sect. 3.4., the energy pattern of an atom–molecule complex, including the transition from the free-rotor limit to a hindered rotor to a rigid asymmetric top, was discussed. We could see that with increasing anisotropy of the potential, the $j = 1$ levels (j describing the CO rotation within the potential) split into two sets of levels: one corresponding to the so-called out-of-plane rotation of the CO monomer ($j = 1$ Π states) and one corresponding to the so-called in-plane rotation of the CO monomer ($j = 1$ Σ states). The Π state resembles an asymmetric-top $K = 1$ state in the rigid-rotor limit ($V_2 \rightarrow \infty$), whereas the Σ state corresponds to a vibration. The energy of the $j = 1$ Π state was determined by McKellar et al. [126]. This rotation is influenced very little by the anisotropy of the potential. The energy separation from the ground state corresponds in the asymmetric-top limit to the rotational A constant and yields information on the angle between the intermolecular and intramolecular axes. For a nearly T-shaped equilibrium structure (as is the case for Ar–CO) the angular dependence can be described by $< 1/\sin^2 \theta >$. The angular dependence is averaged over the particular wave

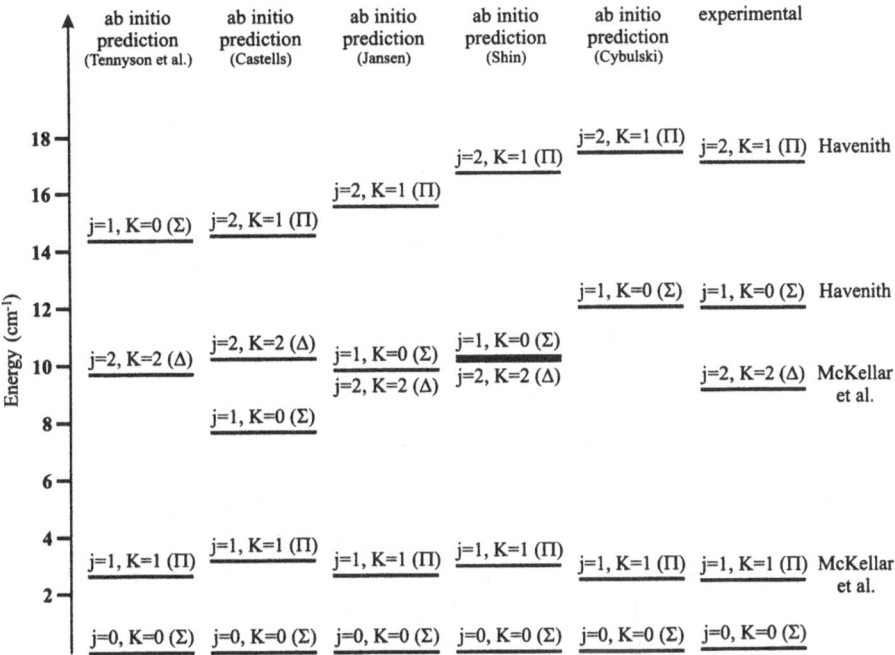

Fig. 8.1. Energy-level scheme comparing the various theoretical predictions for the lowest energy levels with the measured values; see also Tennyson et al. [186], Jansen [96], Castells [26], Cybulski [187], McKellar et al. [126] and Havenith [74, 99]

function describing the motion within the intermolecular potential. The final value depends on the particular intermolecular potential. However, even for large-amplitude motions ($\theta_e = 90°$), the function $1/\sin^2\theta$ depends only weakly on θ, which implies that this energy separation is rather insensitive to changes in the angular anisotropy.

In Fig. 8.1, the experimental data for these levels are compared with the results of several ab initio predictions. We can see that all give rather satisfactory predictions for this energy level.

In the intermediate case of a hindered rotor, the $j = 1$ Σ state can be described as a bending excitation of the CO within the Ar–CO potential. The splitting between the two $j = 1$ stacks depends directly on the anisotropy of the potential, as was shown in Sect. 3.4. More specifically, the energy of this state ($j = 1$ Σ) depends very much on the exact form of the potential and provides a good test of the calculated potentials. This is shown in Fig. 8.1. The energies for the bending state ($j = 1$, $K = 0$, Σ) change considerably for the different potentials: E.g. 7.8 cm^{-1} (Castells), 9.94 cm^{-1} (Jansen) and 14.1 cm^{-1} (Tennyson). The first experimental information on this state became available in 1994, when the transition from the ground state to the $v_{CO} = 1$, $j = 1$, $K = 0$ (Σ) state was measured using our diode laser spectrometer. The details of the measurement are described by Havenith et

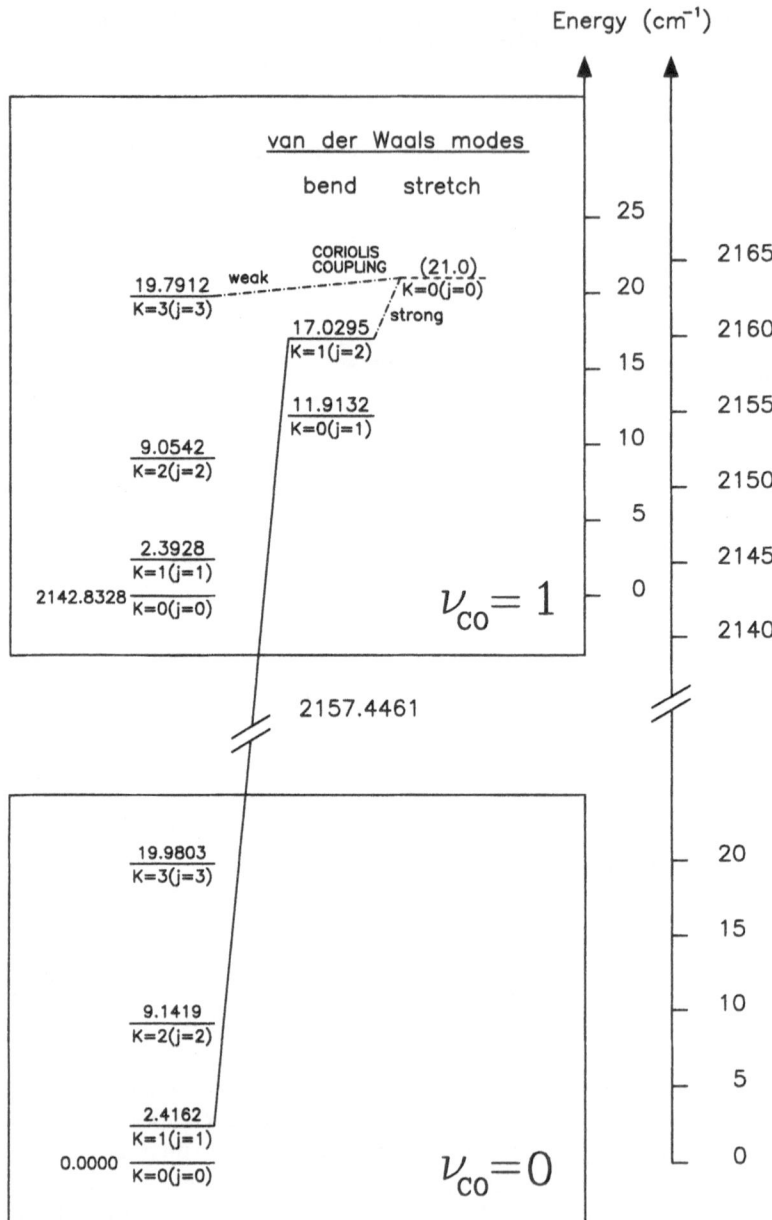

Fig. 8.2. Energy-level diagram showing the sub-band measured by König et al.

al. [74]. The excited bending state was measured on top of the CO vibration, as is indicated in the energy-level diagram in Fig. 8.2.

These measurements yielded an energy difference of 11.91 cm^{-1} from the lowest energy level. Owing to the negligible coupling between the CO vibration and van der Waals modes, this value was considered to be nearly independent of v_{CO}. (Later measurements revealed a value of 12.01 cm^{-1} for $v_{CO} = 0$ [202].) We can see in Fig. 8.1 that none of the theoretical calculations which were available in 1994 were very close to the experimental value. The splitting between the two $j = 1$ levels could be extracted for the first time by the experimental work and amounts to 9.5 cm^{-1}. In comparison, the calculations by Tennyson et al. yielded 11.5 cm^{-1} and the predictions by Castells et al. 4.8 cm^{-1}, clearly overestimating or underestimating the anisotropy of the potential. This new experimental information on the anisotropy was used by Jansen to adjust his pure ab initio potential. The anisotropy of his coupled pair functional (CPF) potential turned out to give the best result for the experimental bending frequency. However, the well depth in the CPF potential is known to be too low. By comparing coupled-pair functional interaction potential surfaces for Ne$_2$, NeAr and Ar$_2$ with empirical potential-energy curves, an extrapolation scheme for the differential correlation energy was suggested. In addition, the anisotropy was modified to reproduce the experimental data [96] as well as possible. This yielded an extrapolated CPF (ECPF) surface with D_e of 109 cm^{-1} at $R_e = 3.68$ Å. Its anisotropy is characterized by a barrier of 19.9 cm^{-1} for rotation around the oxygen end of CO and a barrier of 26 cm^{-1} for rotation around the carbon end. The bending frequency is 11.94 cm^{-1} and the splitting between the two $j = 1$ levels is 9.4 cm^{-1}, which demonstrates how well the potential was adjusted to the experimental data.

The potentials are displayed in Fig. 8.3. The binding energy is plotted as a function of the intermolecular distance (R) and the angle between the intermolecular and intramolecular axes (θ). The potential minimum is found at a distance of $R = 3.68$ Å and an angle of $\theta_e = 97°$. The potential is very flat relative to the angular coordinate. Considering only a variation of R at fixed θ, the potential is more confined. In the region of the repulsive wall, the energy increases considerably with decreasing R. The ground state in this potential has an energy of -85 cm^{-1}. We can see from the potential that even in the ground state, large-amplitude motions are expected. The ground state wave function is expected to show a non-negligible amplitude over a wide range of angles. The bending level at -73.05 cm^{-1} already lies above the barriers for the rotation of the CO monomer within the Ar–CO complex. For a $j = 1$ excitation, the CO monomer can rotate completely (360°) in the plane. Ar–CO is an intermediate case (a hindered rotor); however, it is expected to approach more closely the case of a free rotor than the case of a rigid molecule.

These results show that it is impossible, even for the ground state, to talk of a rigid structure of the Ar–CO complex. However, it explains the broad range of angles θ which were found by Ogata et al. in their deduction of the structure. For the excited bending states, we are above the barrier and will therefore probe the potential over a broad range of angles.

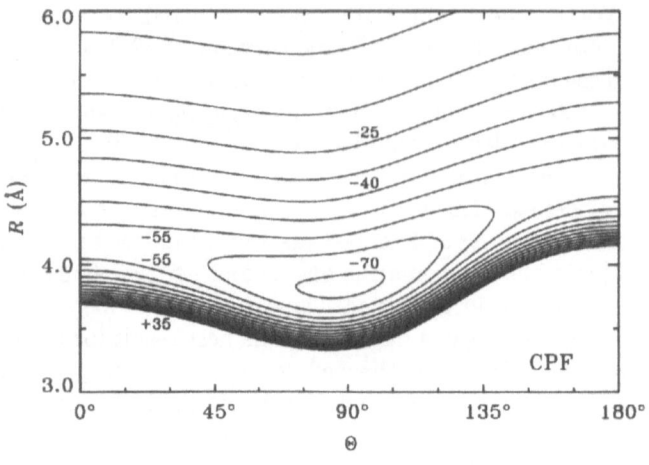

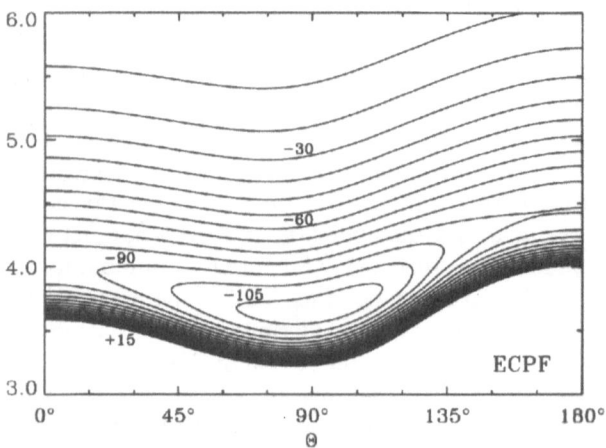

Fig. 8.3. CPF and ECPF potential surfaces of Ar–CO obtained by Jansen ([96])

The most recent ab initio potential surface by Toczylowski and Cybulski [187] is displayed in Fig. 8.4 and resembles very much the main features of the potential of Jansen. The well depth amounts to -104.68 cm^{-1} (-476.97 μ E$_h$), the minimum was found at 93°.

The ECPF potential surface has been used for the assignment of higher van der Waals states. Before we go on to a more detailed description of these measurements, we have to introduce the concept of Coriolis coupling.

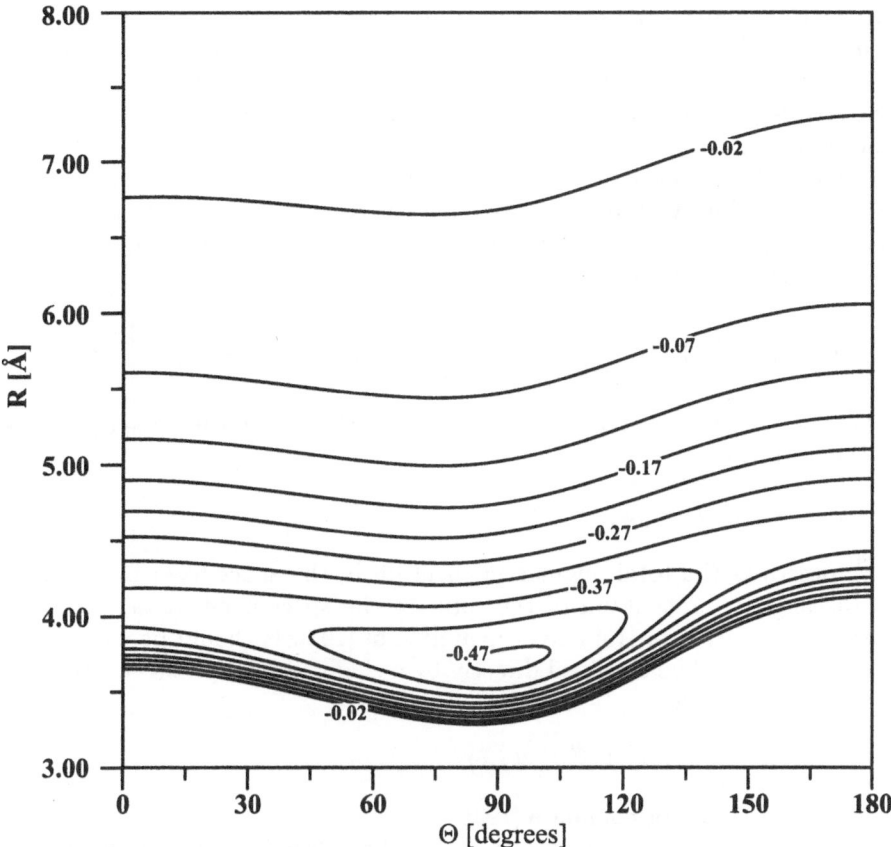

Fig. 8.4. Ab initio potential by Toczylowski and Cybulski ([187]) of Ar–CO. Interaction energies are given in mE_h

8.4 Coriolis Coupling

If we go to higher energies, the density of states will increase. If two states are separated by a small energy gap, they can interact with each other. This can lead to an observable splitting, the Coriolis splitting. For a quantitative discussion of this splitting we have to come back to the Hamiltonian, as described in Sect. 3.4:

$$H_0 = E_{\text{kin}} + V(R, \theta) \ . \tag{8.1}$$

The kinetic energy relative to axes fixed in space (in the "space-fixed" system) can now be written as

$$T = T_A + \frac{\hbar^2}{2\mu R^2} \left(-\frac{\partial}{\partial R} R^2 \frac{\partial}{\partial R} + l^2 \right) , \tag{8.2}$$

where T_A describes the internal kinetic energy of the monomer A, μ is the reduced mass, and l describes the end-over-end rotation of the overall complex. However, although the expression for the kinetic-energy operator in the space-fixed axis system is very convenient, as is demonstrated in (8.2), the potential naturally depends on the internal angles of the complex, i.e. on R and θ, which are defined in the body-fixed axis system. Moreover, T_A can be described within the body-fixed (or complex) axis system easily by $T_A = B(j^{BF})^2$. In the complex embedded frame, the kinetic-energy operator can be written as [193]

$$T = b_{CO}(j^{BF})^2 + \frac{\hbar^2}{2\mu R^2}\left[-\frac{\partial}{\partial R}R^2\frac{\partial}{\partial R} + (J^{SF})^2 + j^2 - 2(jJ)\right], \qquad (8.3)$$

where J is the total angular momentum of the dimer with respect to the embedded frame,

$$J = j + l; \; J_z = j_z \,,$$

J^{SF} describes the total angular momentum in the space-fixed frame and j is the angular momentum of the rotor in the space-fixed frame. An exact derivation is given in [15] and Appendix 4 of [193]. It should be emphasized that (8.3) *cannot* be obtained directly by a change to body-fixed coordinates and the substitution

$$l^2 = (J - j)^2 = J^2 + j^2 - 2(jJ) \,,$$

since j and J do not commute [193].

The basis functions required to solve the Schrödinger equation were introduced in Sect. 3.4:

$$\psi^{JM}(R,\tilde{r},\tilde{R}) = \chi_n(r) \sum_{m_l,m_j} \langle jm_jlm_l|JM\rangle Y_{m_j}^j(\tilde{r})Y_{m_l}^l(\tilde{R}) \,.$$

It can be shown that these are simultaneous eigenfunctions of $(j^{BF})^2$ and $(j^{SF})^2$, with $j(j + 1)$ being the eigenvalue. $|JM\rangle$ are eigenfunctions of the total angular momentum $(J^{SF})^2$ and its projection J_z^{SF}, with the eigenvalues $J(J+1)$ and M [193]. Each state which is in the free rotor limit described by the rotational quantum number j will split into several states (described by J) due to the anisotropic part of the potential. The only rigorous total quantum numbers are the total angular momentum J, its projection M and the parity [193]. The quantum number K, which corresponds to the projection of J onto the intermolecular axis in the complex embedded frame, is nearly a conserved quantum number. This follows from (8.3), since $J_z = j_z$ commutes with the whole Hamiltonian, except for one term which is called the Coriolis term:

$$-\frac{\hbar^2}{2\mu R^2}2(jJ) \,.$$

This can also be written in terms of ladder operators $j^{\pm}$ and pseudoladder operators $J^{\pm}$ as

$$H_C = -\frac{\hbar^2}{2\mu R^2}\left(2j_z J_z + j_+ J_+ + j_- J_-\right).$$

(8.4)

This term will connect eigenfunctions with equal J and j values, but with different K values. In the embedded frame, the following basis functions are used:

$$|n, j, K, p, J, M\rangle = \frac{1}{\sqrt{2}}\,\chi_n(R)\left(\frac{2J+1}{4\pi}\right)^{1/2}$$
$$\times \left[Y_K^j(\tilde{r})D_{M,K}^J(\alpha,\beta,0)^* + pY_{-K}^j(\tilde{r})D_{M,-K}^J(\alpha,\beta,0)^*\right],$$

(8.5)

where p is the parity ($p = \pm 1$). In the case of $K = 0$, both terms in the sum are equal and the factor $1/\sqrt{2}$ has to be replaced by $1/2$. Matrix elements between different basis functions with different K values can be obtained for $K' = K \pm 1$. The connecting matrix elements are given generally by

$$E_{K,K'}^{\text{Coriolis}} = \langle n, j, K, p, M|H_C|n, j', K', p', M\rangle$$
$$= -c\frac{\hbar^2}{2\mu R^2}\delta_{j,j'}\delta_{p,p'}\delta_{K,K'\pm 1}$$
$$\times \sqrt{j(j+1) - K(K\pm 1)}\sqrt{J(J+1) - K(K\pm 1)}\,,$$

where c is a constant, which is $\sqrt{2}$ for the interaction of $K = 0$ with $K = 1$ states and 1 for all other cases. We can further see that only states having the same parity, the same J and the same j can interact with each other; the last two conditions are more rigorous, since j is not a good quantum number owing to the anisotropy of the potential. We shall see in the following section that the Coriolis coupling can lead to an observable splitting in the infrared spectrum of the complex. It therefore has to be included in the analysis of the spectrum.

In the following we shall apply the general concept of the Coriolis coupling to the more specific case of a Coriolis coupling between a $K = 0$ and a $K = 1$ state, since we shall need the results in the next section.

Qualitatively, one nearby state of a specific parity (the $K = 0$ state) can interact with one of the previously near-degenerate $K = 1$ states. This interaction leads to a repulsion between the two interacting states (the $K = 0$ state and *one* parity component of the $K = 1$ state). It is the lower-asymmetry component of the $K = 1$ excited bending state which has the right parity to interact repulsively with the higher-lying $K = 0$ state. This leads to an increased splitting between the two asymmetry components, since the other parity components are not influenced by this nearby $K = 0$ state and remain unchanged. Quantitatively, this has to be described in terms of

Coriolis coupling. The unperturbed component f can be simply described by

$$E_f^0(J) = E_b^0 + B_b^0 J(J+1) - D_b[J(J+1)]^2 . \tag{8.6}$$

For a description of the two interacting states, a (2×2) matrix is used, with the energies of the unperturbed states $E_e^0(J)$ and $E_{K=0}^0$ being the diagonal matrix elements. The Coriolis interaction connects states with different K values and is therefore nondiagonal. The energy of the perturbed states (E) can be found after solving the following matrix equation:

$$\begin{pmatrix} E_e^0(J) & E_{K=0,K=1}^{Coriolis}(J) \\ E_{K=1,K=0}^{Coriolis}(J) & E_{K=0}^0(J) \end{pmatrix} = E \begin{pmatrix} 1 & 0 \\ 0 & 1 \end{pmatrix} . \tag{8.7}$$

According to (8.6), $E^{Coriolis}$ is given by

$$\begin{aligned} E_{K=1,K=0}^{Coriolis}(J) &= -c\frac{\hbar^2}{2\mu R^2}\sqrt{2j(j+1)}\sqrt{J(J+1)} \\ &= -B_{ArCO}\sqrt{2j(j+1)}\sqrt{J(J+1)} \\ &= -C_{Coriolis}\sqrt{J(J+1)} , \end{aligned} \tag{8.8}$$

where $C_{Coriolis} = B_{ArCO}\sqrt{2j(j+1)}$.
$E_{K=0}^0(J)$ and $E_e^0(J)$ are given by

$$E_{K=0}^0(J) = E_{K=0} + B_{K=0}J(J+1) - D_{K=0}[J(J+1)]^2 , \tag{8.9}$$
$$E_e^0(J) = E_b^0 + B_b^0 J(J+1) - D_b[J(J+1)]^2 . \tag{8.10}$$

The energy of the perturbed states can be found from

$$\begin{aligned} E(J) &= \frac{E_b^0(J) + E_{K=0}(J)}{2} \\ &\pm \left(\frac{[E_b^0(J) - E_{K=0}(J)]^2}{4} + \left[E_{K=1,K=0}^{Coriolis}(J)\right]^2 \right)^{1/2} . \end{aligned} \tag{8.11}$$

A full analysis of the Coriolis coupling can be given if both states, i.e. the $K = 0$ state and the $K = 1$ state, are included in the analysis. This was done, for example, by Lovejoy and Nesbitt [118] and Reeve et al. [160] in their analysis of Ar–HCl. From this analysis, the coupling constant $C_{Coriolis}$ can be obtained, which provides direct information on the quantum number j of the states involved. However, since j is not a good quantum number, the j value obtained does not have to be an integer. Owing to the anisotropy of the potential, the eigenstates are linear combinations of basis functions of different j.

However, as long as only one of the two interacting states is known experimentally, a full analysis cannot be performed. Nevertheless, if certain approximations are made, meaningful estimates concerning $C_{Coriolis}$ and the energy difference between the two states $(\Delta E = E_{K=0}^0 - E_b^0)$ can be made. The concept is as follows.

Assuming that the rotational constants of the $K = 0$ state are similar to those describing the unperturbed excited $K = 1$ bending state ($B_{K=0} \approx B_b^0; D_{K=0} \approx D_b^0$), and using (8.11), we obtain the following equation:

$$E_e(J) - E_b^0 - BJ(J+1) + D[J(J+1)]^2$$
$$= \frac{\Delta E}{2} - \left[\frac{(\Delta E)^2}{4} + C_{\text{Coriolis}}^2 J(J+1)\right]^{1/2}. \tag{8.12}$$

Since B, D and E_b^0 are known from a fit of the unperturbed f component, ΔE and C_{Coriolis} can be extracted when the perturbed e component is measured and its energy determined as $E_e(J)$. The two constants C_{Coriolis} and ΔE are uncorrelated, since they show a different J dependence (see (8.12)).

8.5 Measurement of Higher Excited Levels

The measurement of the higher excited levels of Ar–CO gives direct access to higher parts of the potential surface. It therefore provides a sensitive test for the repulsive wall. In the following we shall summarize several measurements which are discussed in detail in the literature [166, 167]. We shall start with the measurement of the first $K = 1$ excited bending level [99]. In the course of these measurements, 47 new lines of the parallel transition from the $K = 1$ ($v_{\text{CO}} = 0$) ground state to the first $K = 1$ ($v_{\text{CO}} = 1$) excited bending level were detected and analyzed. These measurements were motivated by the intensity predictions of Castells et al., who calculated this band to be of comparable intensity to the previously observed transitions from the $K : 0 \leftarrow 0$ band to the first excited bending level.

Parts of the P and R branches of this band are shown in Fig. 8.5. The lower trace displays the reference etalon signal for wavelength calibration. Absolute calibration was performed by reference to the CO lines in the spectrum. These can be easily distinguished from the Ar–CO lines by their broader Doppler profile. (CO is abundant in the whole chamber, which is at room temperature, whereas Ar–CO is only formed in the expansion at 12 K.) The $K = 1$ excited bending state can be described as a simultaneous excitation of an in-plane and out-of-plane rotation/bending of the CO molecule in the Ar–CO potential. The corresponding quantum numbers are $j = 2$, $K = 1(\Pi)$. This state provides further information on the anisotropy of the potential. Since it is a combination band of bending and out-of-plane rotation it will be higher in energy than the bending band. The energy for the $v_{\text{CO}} = 1$ state was determined to be 17.0 cm^{-1}. This coincides very well with the prediction of 16.9 cm^{-1} from the adjusted ECPF potential by Jansen [96], 17.5 cm^{-1} by Shin [176] and 17.4 cm^{-1} by Toczylowski and Cybulski [187].

In the excited state, an unusually large splitting between the two otherwise nearly degenerate energy levels belonging to the same J level was observed. If Ar–CO is considered as a rigid molecule, the splitting between the two

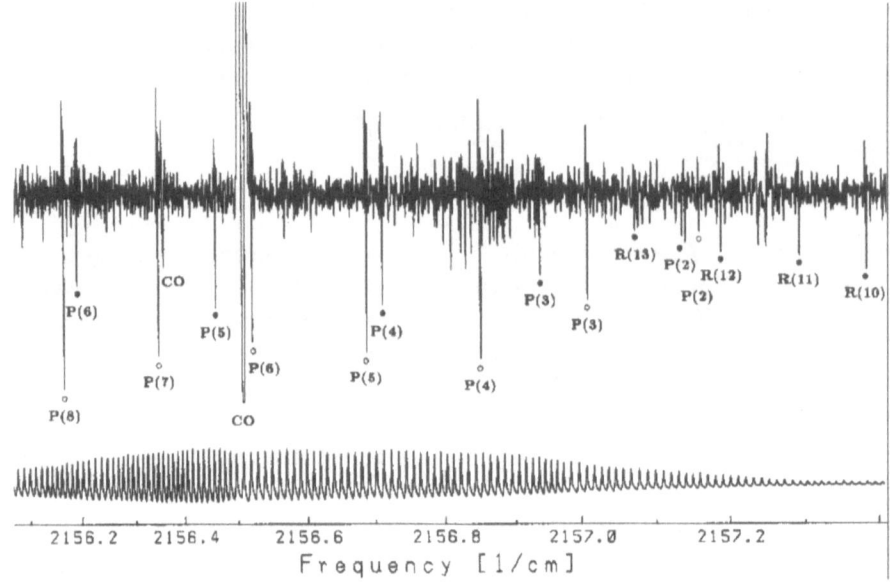

Fig. 8.5. The transition from the $K = 1$ ground state to the $K = 1$ ($v_{CO} = 0$) excited bending level. P(2)–P(8) connecting the upper asymmetry components, are marked by $\circ$; P(2)–P(6) and R(10)–R(12), connecting the lower asymmetry components, are marked by $\bullet$. The Q branch of the perpendicular $K: 4 \leftarrow 3$ transition can be seen around 2156.8 cm^{-1}

different parities is caused by the deviation of the structure from a symmetric top (asymmetry splitting) and reflects the structure. This phenomenon is called K-type doubling. The striking increase of the splitting is demonstrated in Fig. 8.6, in which the splitting of the $K = 1$ excited bending level is compared with the splitting in the $K = 1$ ground state.

For Ar–CO, an explanation of the large splitting as asymmetry splitting is not adequate. However, the splitting can be explained by the concept of Coriolis coupling between nearby states, as discussed in Sect. 8.4.

It was found that the e-component was perturbed by a $K = 0$ state, which lies higher in energy. In an initial step we inserted the rotational constants of the unperturbed state: $B = B_b^0, D = D_b^0$. This yielded the following result: $\Delta E = (4.3 \pm 0.3)$ cm^{-1}, $C_{\text{Coriolis}} = (0.254 \pm 0.007)$ cm^{-1}. To give a more realistic estimate, the calculation was repeated with a slightly reduced B constant. As the perturbing state could be identified as the van der Waals stretch (see below), the value of $B_{K=0}^0$ is expected to be smaller than B_b^0. Reducing B by 7.5%, which corresponds to a realistic estimate for the stretch on the basis of experience with Ar–HCl, the following predictions were obtained:

$$\Delta E = (1.45 \pm 0.07)\, \text{cm}^{-1}, \quad C_{\text{Coriolis}} = (0.122 \pm 0.002)\, \text{cm}^{-1}.$$

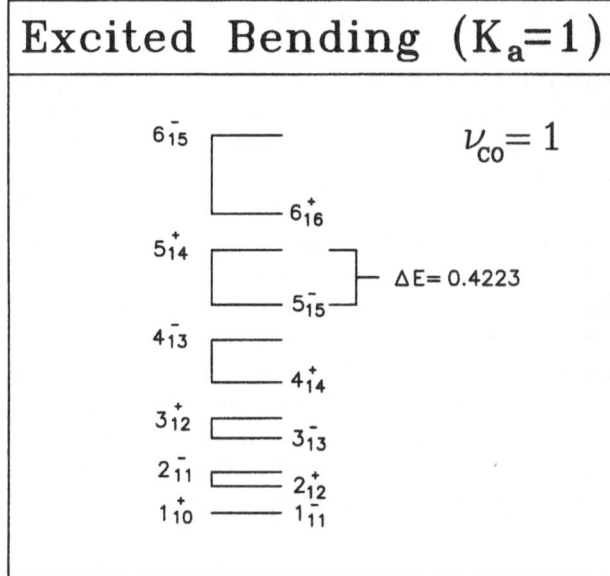

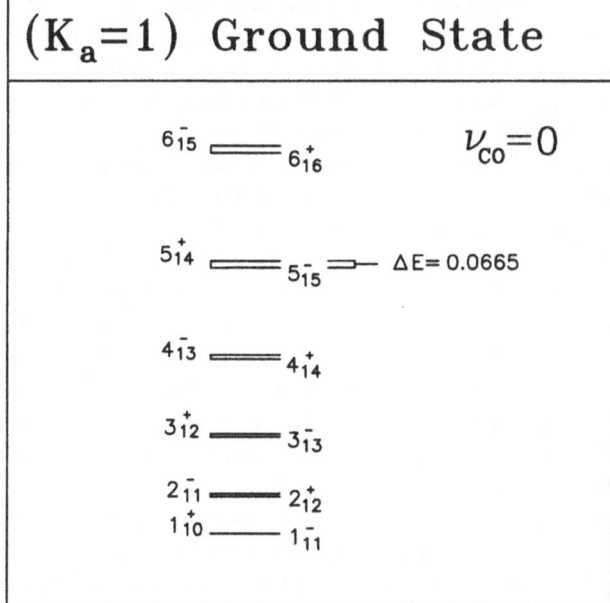

Fig. 8.6. Energy levels of Ar–CO showing the increase in the asymmetry splitting for the $K = 1$ excited bending level compared with the $K = 1$ ground state. The levels are characterized by $J^p_{K_a,K_c}$

This demonstrates that 0.254 cm^{-1} is an upper limit, with the second estimate giving a more realistic estimate of the true values.

Considering the theoretical predictions the perturbing $K = 0$ state was identified as the van der Waals stretch. This state is, more precisely, a mixture of the ($j = 2$, $K = 0$) doubly excited bending state and the $j = 0$ van der Waals stretch, with more $j = 0$ character. For a pure $j = 2$ state, a coupling constant of $B\sqrt{j(j+1)} = 0.23$ cm^{-1} ($B = 0.067$ cm^{-1}) for the coupling to the $j = 2$ bending state is predicted. Assuming a pure van der Waals stretch ($j = 0$), the coupling should be zero, since different j values cannot interact by Coriolis coupling (see (8.6)). The above estimate of 1.45 cm^{-1} is therefore in good agreement with the assumption that the perturbing state is a mixture of a $j = 0$ and a $j = 2$ state. The energy difference as predicted by Jansen ($\Delta E = 1.4$ cm^{-1}) is close to the values estimated above, which clearly supports the above assignment. In subsequent measurements by Xu and McKellar the perturbing state could be directly observed [202]. The exact analysis is in very good agreement with the estimates obtained from the approximation and yields $\Delta E = 1$ cm^{-1}, $C_{\text{Coriolis}} = 0.131$ cm^{-1}. What can we learn from these numbers about the potential? The stretch ($j = 0$, $K = 0$) obviously contains a non-negligible contribution of the doubly excited bend ($j = 2$, $K = 0$), which is caused by the anisotropy of the potential mixing equal K and different j values. This leads to relatively large Coriolis coupling constants for Ar–CO, which exceed the coupling constants known for Ar–HCl ($C_{\text{Coriolis}}^{\text{Ar–HCl}} = (0.04–0.09)$ cm^{-1} [118]). A comparison between the different potentials reveals that the Ar–CO potential is more anisotropic than the Ar–HCl potential. This implies that the variation of R with θ for the minimum-energy tunneling path is significantly larger in Ar–CO than in Ar–HCl.

Using a Herriott multipass cell, which increased the sensitivity by a factor of 10, it was possible to detect several transitions from even higher levels. The measurement of these lines, which are about a factor of 10 smaller than the lines discussed before, is described in detail by König and Havenith [100] and by Scheele, Lehnig and Havenith [166, 167]. In 1997 König et al. detected a sub-band which could be assigned as a K: 1 ← 2 transition. The ground state coincides with the $j = 2$, $K = 2$ out-of-plane rotation as found by McKellar et al.[126]. The upper level was identified as a combination band of the stretch and the out-of-plane rotation ($j = 1$, $K = 1$) with $v_{CO} = 1$, since this band was the only $K = 1$ state which was predicted to have an energy close to the observed energy of 26.2 cm^{-1}.

The R branch was observed to be significantly weaker than the P branch. This supports our statement that Ar–CO is no longer in the rigid-rotor limit (where the P and R branches have equal intensity) but has a hindered-rotor characteristic. For a free rotor, the selection rules $\Delta J = 0, \pm 1$; $\Delta j = \pm 1$, $\Delta l = 0$ would apply, since the transition dipole moment of the complex is carried exclusively by the CO molecule. If we look at the correspondence between these quantum numbers and the J, K quantum numbers used here, we see (e.g. Fig. 3.3, Sect. 3.4) that for a free-rotor $\Delta K = -1$ sub-band only the P branch would fulfill the selection rules. In the case of a hindered rotor, the P branch has an increased intensity compared with the R branch, which is completely forbidden in the free-rotor case.

The fit of the $K_a = 1$ state at 26.187 cm^{-1} yielded a rotational constant of 0.073 cm^{-1} for the upper state, which is unusually large for a stretching mode. (For comparison, Xu and McKellar found a B-value of 0.0618 cm^{-1} for the stretching fundamental mode [202] .) The asymmetry splitting is found to be slightly smaller than in other $K_a = 1$ states. This indicates the presence of Coriolis coupling to a low-lying $K_a = 0$ state.

In 2000 the perturbing state was directly observed by Scheele, Lehnig and Havenith [166, 167]. A $K_a = 0$ at 23.927 cm^{-1}was reported, which was assigned to a combination band of the stretching and bending mode.

For a complete analysis the two sub-bands were fitted simultaneously, taking Coriolis coupling between these two states explicitly into account. The energies of the two coupling states were obtained by diagonalizing the 2×2 matrix given in (8.7).

The decoupling leads to an effective B-rotational constant of 0.0709 cm^{-1} which is decreased compared to the previous value but still higher than the expected value (e.g. $B_{\text{stretch}} = 0.062$ cm^{-1} [202]). This indicates that the $K_a = 1$ state is still affected by Coriolis coupling to the $K_a = 2$ bending state which is predicted to lie close but which has not been observed yet. Since the rotational constant is still too high we expect a Coriolis coupling with a lower-lying $K_a = 2$ state. This state is estimated to lie below the $K_a = 0$ state around 23 cm^{-1}.

It is striking that none of the best and very recent ab initio calculations is able to predict the energies of these states to a satisfying level. Whereas all theoretical studies predict the $K_a = 0$ state above the $K_a = 1$ state the experiments clearly reveal the $K_a = 0$ combination band of the stretching and bending mode to lie below the $K_a = 1$ state of the stretching fundamental. This disagreement indicates the lack of some important feature in the potential energy surface.

The highest observed van der Waals mode in the intermolecular potential energy surface was reported by Scheele, Lehnig and Havenith [167]. Whereas for the higher states a more chaotic behavior was expected, a $K_a = 0$ state at 36.765 cm^{-1}showed a quite regular pattern and no indication of any perturbation. It seems that the Ar–CO potential which is dominated by the balance of dispersion and repulsion interaction can serve in the future as a benchmark for the description of intermolecular forces. The experimental data available shall serve as a sensitive test for future ab initio calculations.

In order to be able to guide further theoretical work we performed a preliminary semi-empirical fit of the potential energy surface.

8.6 Semi-empirical Fit of the Potential Surface

The measurement of the higher excited states showed that there was still
a need for improvement of the theoretical potential surfaces. We therefore
performed a semi-empirical fit of the potential surface of Ar–CO using the
experimental results described in the previous sections. The proper form of a
semi-empirical potential is the subject of Chap. 3. Whereas the exact details
of the fit referred to will be subject of a forthcoming paper [161], only the
main results will be summarized here. Since the number of parameters in
the potential surface which can be fitted is limited, we have to start with a
realistic ab initio potential, in which only a few contributions, such as the
dispersion interaction and the repulsion, are modified, in order to obtain a
good agreement with the experimental observations.

The adjusted ab initio potential of Jansen should be a good starting point
for further improvements, since it yielded a good agreement for the lower van
der Waals states and a considerable improvement over the previous ab initio calculations, which failed to provide reliable predictions of the van der
Waals states (see Fig. 8.7). The first accurate pure ab initio calculations of
Ar–CO known were published by Shin et al. in 1996 [176]. This calculation
lacks any consideration of Coriolis coupling, which makes it less helpful for
the interpretation of the experimental data, since the inclusion of Coriolis
coupling has turned out to be an important concept. Nevertheless, the potential of Shin et al. can be used as an initial potential for a semi-empirical
fit of the potential surface. In the first step towards the establishment of a
semi-empirical potential, the potential surfaces of Jansen and Shin, which
were calculated pointwise on grids, had to be parameterized. For an accurate
description, the following functional form was chosen:

$$V(r, \theta) = A(\theta) \exp\left[-b(\theta)R - b_2(\theta)R^2\right] + \sum_n \frac{C_n(\theta)}{R^n} f_n[b(\theta, R)] . \qquad (8.13)$$

This description of the potential is similar to the functional form introduced
in Chap. 3. It includes a repulsive contribution which is more flexible than
the pure exponential used in Chap. 3. It turns out that the introduction
of the term $\exp\left[-b_2(\theta)R^2\right]$ is necessary for an accurate description of the
SCF part of the MP4 potential of Shin et al. For an accurate description
of the dispersion, an additional exponential term has to be introduced. This
represents higher-order corrections to the dispersion such as the exchange–
dispersion term, similar to that used in the calculation for He–CO [136]. The
parameters A, b, b_2 and C_n are θ-dependent, which implies that the Tang–
Toennies damping function will also depend on the angle θ. The dispersion
and the induction energy are represented by (see also (3.1))

$$V_{\text{disp+ind}} = \sum_{n=6}^{10} f_n(R) \frac{C_{\text{disp,n}}}{R^n} + \sum_{n=6}^{7} f_n(R) \frac{C_{\text{ind,n}}}{R^n} . \qquad (8.14)$$

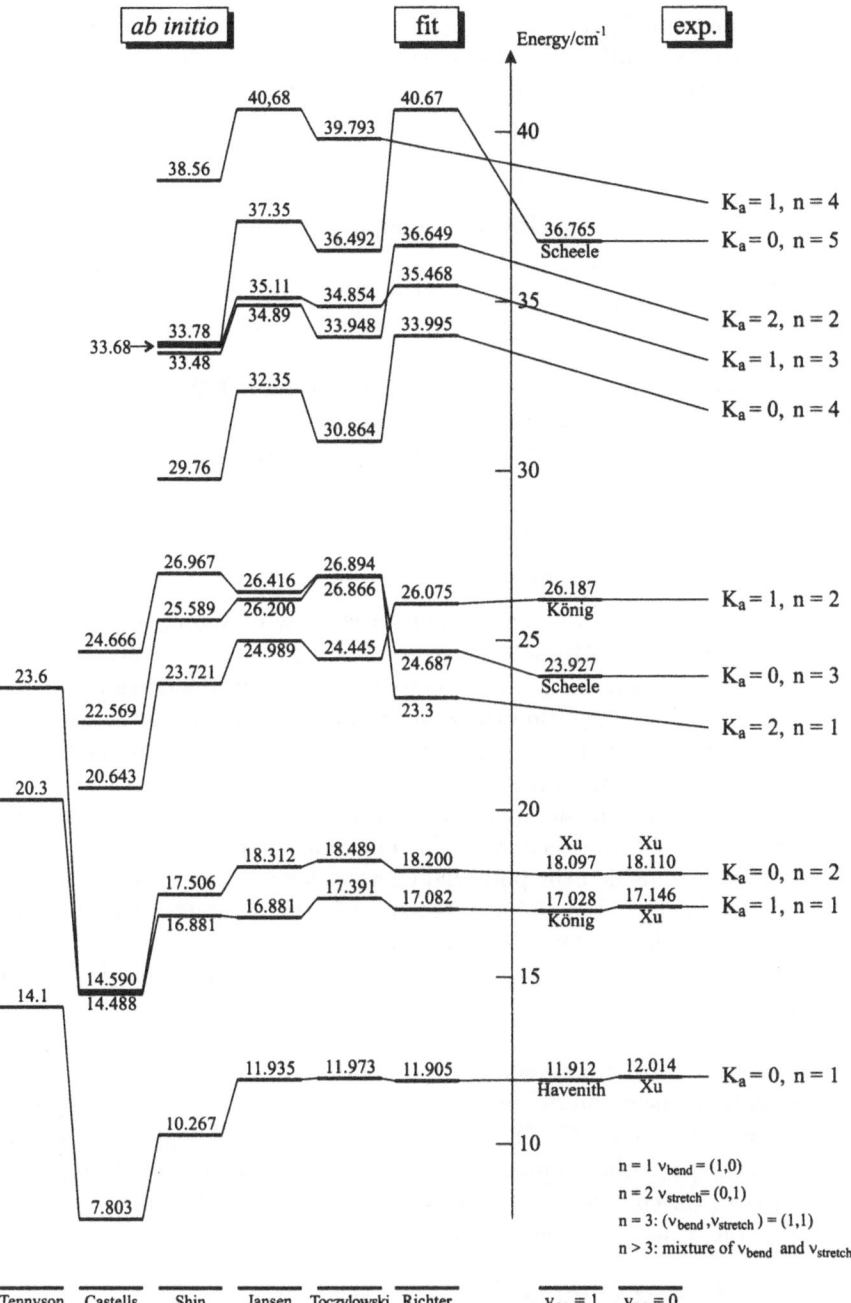

Fig. 8.7. Energy-level scheme showing the van der Waals (bend and stretch) states of Ar–CO. Ab initio calculations by Tennyson [186], Castells [26], Shin et al. [176] and Toczylowski [187], the predictions of the adjusted ab initio potential of Jansen [96], and the results of a semi-empirical potential fit by Richter et al. [161] are compared with the experimental data [74, 99, 100, 166, 167, 202]

The parameters $C_{\text{disp},6}$ to $C_{\text{disp},10}$ for the dispersion energy were taken from ab initio calculations by Haettig and Heß [66]. The leading dispersion coefficients were expressed as linear Legendre expansions, with

$$C_6(\theta) = C_6^0 + C_6^2 P_2[\cos(\theta)] ,$$
$$C_7(\theta) = C_7^1 P_1[\cos(\theta)] + C_7^3 P_3[\cos(\theta)] ,$$

and so on. We used a Tang–Toennies damping function (see Sect. 3.1):

$$f_n(R, \beta) = 1 - \exp(-\beta R) \sum_{k=0}^{n} \frac{(\beta R)^k}{k!} .$$

In a way similar to the treatment in [136], it turned out that the parameter $\beta(\theta)$ could not be set equal to the parameter b of the SCF contribution, but had to be varied independently. For the induction energy, we used the relations [90]

$$C_{\text{ind},6} = -\alpha_{\text{Ar}} \mu_{\text{CO}}^2 [1 + P_2(\cos \theta)] ,$$
$$C_{\text{ind},7} = -6\alpha_{\text{Ar}} \mu_{\text{CO}} \Theta_{\text{CO}} \cos^3(\theta) ,$$

where α_{Ar}, μ_{CO} and Θ_{CO} describe the polarizability of argon, the dipole moment of CO and the quadrupole moment of CO [128], respectively. However, for Ar–CO, the induction energy amounts to only 2% of the overall potential energy. This small contribution turns out to have no substantial influence on the energies of the bound states. The induction energy was therefore fixed in the further course of the work.

It turned out that the number of ab initio energies was not sufficient for a fit to the functional form given in (8.13). Therefore, a spline interpolation was performed, which yielded the potential energies for additional R values. This allowed a reliable fit of the ab initio potential to the parameterized potential. The quality of this fit could be tested by a comparison of the calculated energies for the bound states as reported by Jansen and Shin et al. with the values obtained using our parameterized potential. The agreement turned out to be very good (within 2%).

For variation of the parameters in the potential, we followed the method introduced by Hutson and discussed in Sect. 3.2 of this book: instead of varying the parameters $A(\theta), C_n(\theta)$ and $b(\theta)$ directly, the well depth $E_{\min}(\theta)$, the position of the potential minimum $R_{\min}(\theta)$ and the steepness of the repulsive wall $b(\theta)$ were varied. These new parameters can be directly related to A and C_n, if the following conditions are used:

$$V(R_{\min}, \theta) = -E_{\min}(\theta), \ V'(R_{\min}, \theta) = 0 .$$

The exact relations can be derived directly from these conditions and are given explicitly in [93].

The preliminary results for the semi-empirical fit of the potential surface are given in Tables 8.1 (for $v_{\text{CO}} = 0$) and 8.2 (for $v_{\text{CO}} = 1$). In these tables

Table 8.1. Energy levels of Ar–CO (in cm^{-1}) for $v_{CO} = 0$

Jansen ECPF [96]	Shin MP4 [176]	Cybulski [187]	Semi-empirical fit [161]	Experimental
2.539	2.946	2.45	2.484	2.416
9.434	10.128	–	9.187	9.142
20.464	21.344	–	19.957	19.978
11.935	10.267	11.937	12.008	12.014
16.900	16.685	17.391	17.197	17.144
18.312	17.506	18.489	18.103	18.110

Table 8.2. Energy levels of Ar–CO (in cm^{-1}) for $v_{CO} = 1$ (Theoretical values are restricted to $v_{CO} = 0$)

Semi-empirical fit [161]	Experimental
2.381	2.393
9.092	9.054
19.795	19.791
34.476	34.511
11.905	11.912
17.082	17.026
18.200	18.097
24.987	23.93
25.115	–
26.180	26.187

the experimental data are compared with the results of the various calculations. (In the calculations of Jansen, Shin et al., and Toczylowski et al. the dependence of the potential on v_{CO} was not taken into account.) In the upper half of each table the energy levels of the out-of-plane rotations are listed; in the lower half the energy levels describing the different van der Waals modes (bend/stretch) are given.

For the energy levels up to 20 cm^{-1} the experimentally determined energies of the intermolecular modes are in very good agreement with the theoretical predictions. However, even the most sophisticated ab initio methods are unable to reproduce the higher-lying levels. The semi-empirical fit is able to reproduce the correct energy level sequence. In the region 20–27 cm^{-1} the preliminary semi-empirical fit can therefore be considered as a significant improvement, indicating important features in the potential surface which had been lacking so far. The direct comparison shows the following changes in the potential surface:

- The well depth E_{min} is increased compared to the results of ab initio studies. Whereas a well depth of 110 cm^{-1} was predicted by the ECPF and MP4 potentials the fit yielded a well depth of 127 cm^{-1}. However, it is known that the well depth is highly correlated to the steepness of the potential. The large deviation from the most recent ab initio predictions might therefore be considered as an artefact resulting from of a not fully adjusted fit.

- In order to reproduce the energy levels to the required accuracy, the rotational constant of the CO monomer had to be varied. This was surprising at first, because in other studies the intramolecular structure is assumed to be unaffected by the intermolecular interaction. However, for Ar–CO we find that the rotational constant of CO needs to be reduced by 1.3% (for $v_{CO} = 0$) or 1.8% (for $v_{CO} = 1$) in comparison to the free monomer. This implies an increase in the distance between the carbon and the oxygen atom by 0.7% or 0.9% owing to the presence of the argon atom. This can be explained by repulsion due to the electron cloud of the argon atom.

- Compared to the potential of Jansen, a second local minimum at 180 ° can be found for the semi-empirical potential surface. The first global minimum at 90 ° can be clearly attributed to a minimum in repulsion. The existence of a second minimum is quite interesting. Since this second minimum is 30 cm^{-1} above the potential minimum, it is clear that the existence of this minimum will have a larger impact on higher van der Waals modes than on lower-lying states.

The final potential is shown in Fig. 8.8. This potential surface shows several features which are characteristic of the intermolecular potential of a noble-gas–molecule interaction lacking induction. The potential is dominated by a balance of repulsion and dispersion. In order to minimize the repulsion and have a maximum of dispersion energy, the carbon and oxygen atom are repelled as far as possible from the argon, which yields a perpendicular equilibrium structure. The CO monomer rotates around its center of mass, which is slightly closer to the oxygen than to the carbon. As a consequence, the barrier for rotation is slightly larger when the carbon approaches the argon (13.7 cm^{-1}) than when the oxygen approaches the argon (20.4 cm^{-1}), for a fixed R value, i.e. a fixed argon–(CO) center-of-mass distance. The potential is more anisotropic than the Ar–HF potential. Whereas for Ar–HF the various contributions, namely dispersion, induction and repulsion, add up such that R_{min} stays nearly constant, for Ar–CO we see a different behavior. If θ approaches 0° or 180°, which means that the carbon or the oxygen atom is rotated towards the argon, the repulsion of the argon electron cloud leads to an increase of R_{min}; otherwise, the carbon or the oxygen atom with its electron cloud would come too close to the argon atom. In the case of Ar–HF only the hydrogen atom, with an effective electron charge of less than one, rotates to a first approximation. This gives rise to a much less anisotropic repulsion than for Ar–CO. The anisotropy observed for Ar–CO should therefore be typical of all noble-gas–molecule complexes in which the molecule consists exclusively of heavy atoms. The same effect can also be found in the potential surface of Ar–CO$_2$. The latest results of Hutson [94] show an anisotropic potential-energy surface, in which R_{min} changes from 3.4 Å at 0° to 4.6 Å at 180°. In

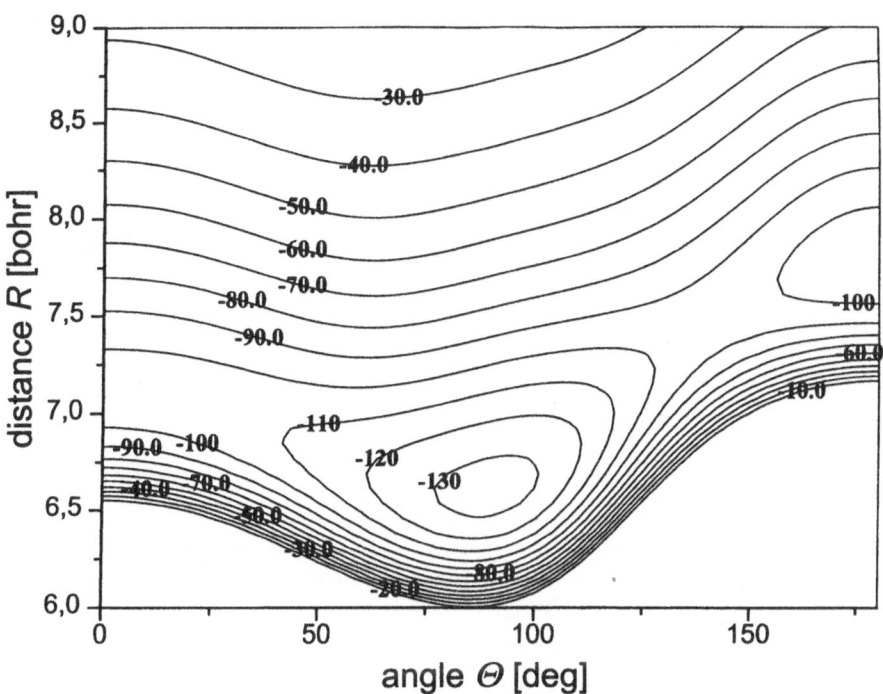

Fig. 8.8. Potential-energy surface as obtained by a preliminary semi-empirical fit of the experimental results

comparison, for Ar–CO $R_{\min}$ varies from 3.7 to 4.4 Å if the CO monomer is rotated from 0° to 180°. The change in $R_{\min}$ is slightly more pronounced for Ar–CO_2 than for Ar–CO owing to the increased monomer size. Owing to the anisotropy, the rotation of the CO monomer or the bending is coupled to the van der Waals stretch. This was clearly observed for Ar–CO and leads to the coupling of the ($j = 2$, $K = 0$) bending state to the ($j = 0$, $K = 0$) stretch, which yields a large Coriolis coupling constant (compared with the Coriolis constants for Ar–HF). We see that the interpretation of the spectra discussed in the previous sections depends on a knowledge of the potential surface. At the same time, the new spectra allow, for the first time, a derivation of the exact form of the intermolecular potential for Ar–CO as a prototype of the intermolecular interaction of a noble-gas–molecule complex, or more specifically of a noble gas interacting with a molecule which consists exclusively of heavy atoms. An improved understanding of this small relatively simple system is required for the development of adequate descriptions of the intermolecular interaction in general.

9. $(NH_3)_2$ – a Prototype
of Hydrogen Bonding?

9.1 Hydrogen Bonding

One of the important concepts in physical chemistry is the hydrogen bond. It is well known that there is a significant attractive interaction between a hydrogen atom attached to an electronegative atom, such as oxygen, a halogen or nitrogen, and another electronegative atom. The strength of this interaction clearly exceeds the values for the normal van der Waals interaction. Binding energies for hydrogen bonding are typically on the order of $1000-3000$ cm^{-1}, whereas van der Waals bonds have energies in the range of $100-300$ cm^{-1}. The relevance of hydrogen bonding is demonstrated by the fact that it is directly responsible for the secondary structure of protein chains and for the duplication of the genetic code. There is still some controversy concerning the nature of hydrogen bonding. For a detailed discussion see, for example, [182]. The hydrogen bond was first considered to be a pure electrostatic effect [151] and only point charges were taken into account. This model failed in many cases, which gave rise to the conclusion that additional charge transfer terms were an important contribution to the total binding energy of a hydrogen bond; see, for example, [159]. It was argued then that an actual amount of electron charge (a net charge) was transferred from the electron donor to the electron acceptor. Effective numbers for the net charge can be found in [159] for various hydrogen-bonded complexes. This charge transfer is assumed to contribute substantially to the stabilization of the bond. Hydrogen bonds would therefore have some contribution from incipient chemical bonding.

However, the failure of a simple point charge model is not surprising in view of the previous discussion in Chap. 2. If only point dipoles are included, as in most cases for these simple models, a linear structure is predicted for $(HF)_2$. But the inclusion of higher-order multipole moments leads to a correct description of the observed structure even when the other contributions, such as induction and dispersion, are neglected. For a correct description – without any charge transfer between the molecules – these contributions have to be included. The question of whether additional charge transfer takes place requires therefore an exact knowledge of these other contributions (electrostatic, induction, dispersion and repulsion), which have nothing to do with real chemical bonding. This is a nontrivial endeavor, and the discussion about charge transfer has still not come to an end. However, it should be mentioned that recent calculations of charge transfer terms in hydrogen bonds yielded the result that their contribution to the overall binding energy is small compared with the other contributions [181]. Furthermore, recent semiempirical calculations of the intermolecular potential, which describe the experimental data very well, do not include any charge transfer term, which might be taken as further evidence that the contribution to the total binding energy was overestimated previously.

A simple, but very successful, model for the prediction of the structure of hydrogen-bonded complexes was postulated in 1983 by Buckingham and Fowler [20]. This so-called Buckingham–Fowler model included the following:

- The electrostatic interaction is taken into account. This includes the consideration of high-order multipole moments and distributed multipole expansions (see Sect. 3.2 and [182]).
- Induction and dispersion are completely neglected. This has the consequence that no meaningful binding energies can be obtained from this model.
- The repulsion is described by a crude hard-sphere model. Each atom, with the exception of hydrogen, is assigned a standard van der Waals radius. The repulsion goes from zero to infinity as soon as the distance is smaller than the van der Waals radius. The hydrogen in the hydrogen bond is assigned a radius of zero.

The success of this model, which correctly predicted many structures [20], lies in the significance of the electrostatic contribution for the determination of the structure for strongly polar complexes. This does not necessarily mean that the electrostatic terms have to provide the main contributions to the binding energy, but only that they provide the most anisotropic contributions. The description of the repulsion is certainly something which could be improved by more accurate models. For this purpose we have tested the contribution of the repulsion in various complexes for which accurate semiempirical potentials are available, e.g. Ar–H$_2$, Ar–HBr, Ar–HCl and Ar–HF. If we compare the repulsion in Ar–HF for $\theta = 0°$ (the hydrogen is pointing towards the argon atom) with the repulsion for $\theta = 180°$ we can clearly see that the repulsion of the hydrogen atom cannot be neglected. If we fix the argon–fluorine distance at R_m and vary θ between 0° and 180°, the repulsion decreases from 257 cm^{-1} to 57 cm^{-1}. If the repulsion of the hydrogen can be neglected, as in the Buckingham–Fowler model, the repulsion should be constant, which is in clear contradiction to the results for the most recent semiempirical potentials. We can, further, state that the repulsion of the hydrogen atom depends on the net charge of the hydrogen in the chemical bond. We have therefore compared the repulsion at $\theta = 0°$ for the various argon–hydride complexes. The result is shown in Fig. 9.1.

The figure shows the repulsion (as a function of the hydrogen–argon distance) at $\theta = 0°$ for various argon–hydride complexes on a logarithmic scale.

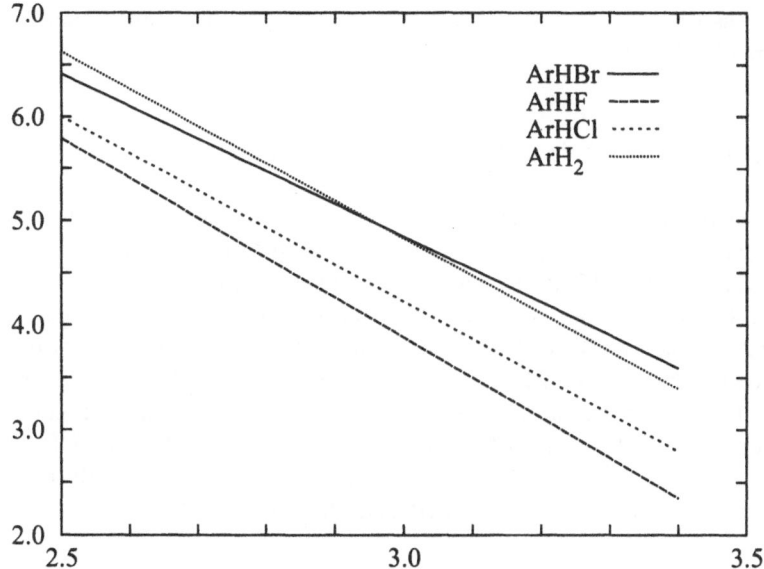

Fig. 9.1. Repulsion (on a logarithmic scale) of the hydrogen atom for various argon–hydride complexes

We can clearly see an increase in repulsion from Ar–HF to Ar–H$_2$. This can be attributed to the increase in the net charge on the hydrogen atom. Owing to the chemical bond, the effective charge left on the hydrogen bond decreases from Ar–H$_2$ (effective charge $1e$) towards Ar–HF (effective charge ca. $0.6e$). If we compare the slopes for small distances (where the hydrogen atom is the main contributor to the repulsion) we can see that this results in an effective decrease of the van der Waals radius by ca. 0.2 Å. An improved model should therefore consider the repulsion of the hydrogen; however, the van der Waals radius of the hydrogen atom should be selected according to the chemical bond in which the hydrogen atom is involved.

Since the charge of the hydrogen atom in a chemical bond is reduced, the reduction of the repulsion of the hydrogen atom in van der Waals bonds is a more general phenomenon. It is not restricted to the small noble-gas complexes discussed above (these were investigated because exact quantitative descriptions of the potential are available), but will certainly be of major importance in hydrogen bonds. The reduction of the repulsion can be considered responsible for a decrease of R and accordingly for an increase in the binding energy E_{tot}, as observed in hydrogen bonds. We shall investigate the role of electrostatic energy and repulsion in more detail in the following sections, which treat the ammonia dimer as a further prototype system for the study of intermolecular forces. The subject of this section will be discussed in more detail in a forthcoming paper [174]. A summary of the ammonia dimer results, as presented in the next section, will be published in a joint paper by all groups involved [194].

9.2 The Ammonia Dimer – An Introduction

The ammonia dimer has been the subject of controversy for the last ten years. The central point in this discussion is the question of whether the well-known ammonia molecule can act as a proton donor or not. There is no doubt that ammonia can serve as a proton acceptor. Several studies of ammonia-containing complexes have proven the latter assumption: ammonia forms hydrogen bonds even with the weakest proton donors. If ammonia can also act as donor, then the ammonia dimer is expected to show a classical hydrogen-bonded structure, as is the case for the analogous dimers of H_2O and HF. By now, it is well established experimentally and theoretically that $(H_2O)_2$ and $(HF)_2$ have nearly linear hydrogen bonds (Fig. 9.2).

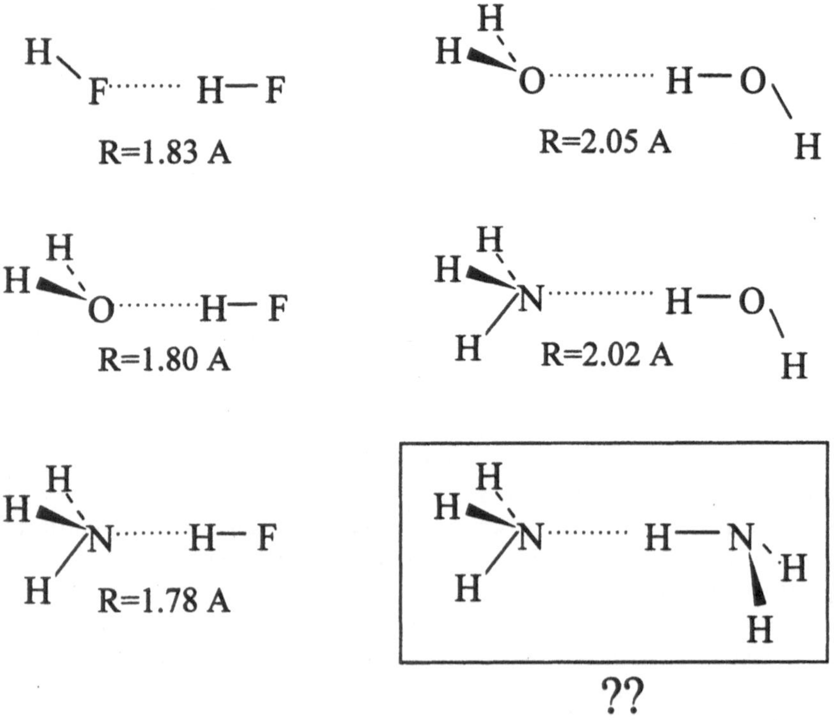

Fig. 9.2. Structure of several H_2O and HF complexes showing the typical nearly linear hydrogen-bonded structure. As a result, $(NH_3)_2$ was also expected to show a linear hydrogen-bonded structure

Until 1985 it was generally assumed that the ammonia dimer also had a classical hydrogen-bonded structure, with the basic monomer oriented with its lone pair of electrons pointing towards the donor hydrogen atom. The most characteristic feature of hydrogen bonding in $(NH_3)_2$ would be a near linearity of the NH bond of the proton donor with the C_3 axis of the proton acceptor. The general chemical intuition that $(NH_3)_2$ would form a hydrogen bond agreed with the results of all ab initio calculations at that time [55, 106, 114]. The expected hydrogen-bonded structure is shown in Fig. 9.2. This predicted structure is displayed in more detail in Fig. 9.3. The intermolecular forces are described by a six-dimensional potential surface: the interaction energy depends on the bond length R and on the angles θ_1, θ_2, ϕ_1, ϕ_2 and γ, where θ_1 and θ_2 are the angles between the threefold C_3 symmetry axes of each NH_3 monomer and the bond vector $\boldsymbol{R}$ (i.e. the dimer a axis). The angle γ is the dihedral angle between the two C_3 axes. The angles ϕ_1 and ϕ_2 describe the rotations of the monomers around their C_3 axes.

In 1985 microwave experiments were carried out in the group of Klemperer [140] at Harvard which initiated a vigorous debate among physical chemists as to whether $(NH_3)_2$ is actually hydrogen-bonded. This debate lasted for about ten years. Recent experimental and theoretical results have now established a higher level of understanding of this prototypical complex; these results are presented below.

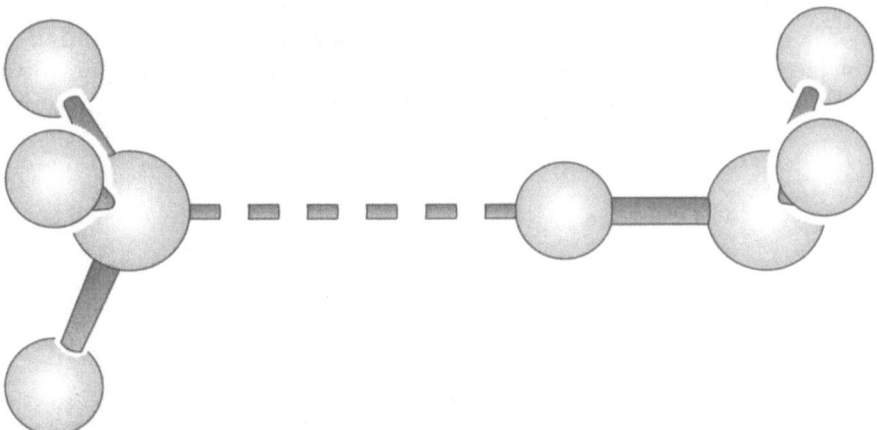

Fig. 9.3. Structure of $(NH_3)_2$. Displayed is the so-called hydrogen-bonded structure, as predicted from previous ab initio calculations

9.3 Previous Results

Nelson, Fraser and Klemperer [140] studied $(NH_3)_2$ by high-resolution microwave spectroscopy using the molecular-beam electric-resonance technique. They observed two sets of apparently rigid-rotor rotational transitions, from which the following quantities could be deduced: the rotational constant describing the overall rotation of the complex, the projection of the electric dipole moment on the a inertial axis of the complex (0.74 D[1]) and the diagonal components of the ^{14}N nuclear electric-quadrupole coupling tensors along this axis. The weak intermolecular forces leave the NH_3 monomer properties nearly unchanged.

The rotational constants are therefore a direct measure of the intermolecular distance between the two NH_3 monomers, and yield $R = 3.34$ Å. Assuming that the dimer is a rigid molecule, the measurement of the molecular dipole moment and ^{14}N nuclear quadrupole couplings provides information about the orientation of the ammonia monomers with respect to the a axis of the complex.

More specifically, one can deduce from these quantities the angles θ_1 and θ_2 between the symmetry axis of each ammonia molecule and the a axis of the complex. The structure that was obtained by Nelson et al. [140] is displayed in Fig. 9.4 and will, in the following, be called the "cyclic" structure. The NH_3 monomers are aligned nearly antiparallel and their polar angles are $\theta_1 = 49°$ and $\theta_2 = 65°$.

The rigidity of the two states observed was attributed to the dynamical inequivalence of the monomeric units, i.e. ortho–para and para–ortho combinations. States formed from identical monomers, i.e. para–para and ortho–ortho, were assumed to not have been observed.

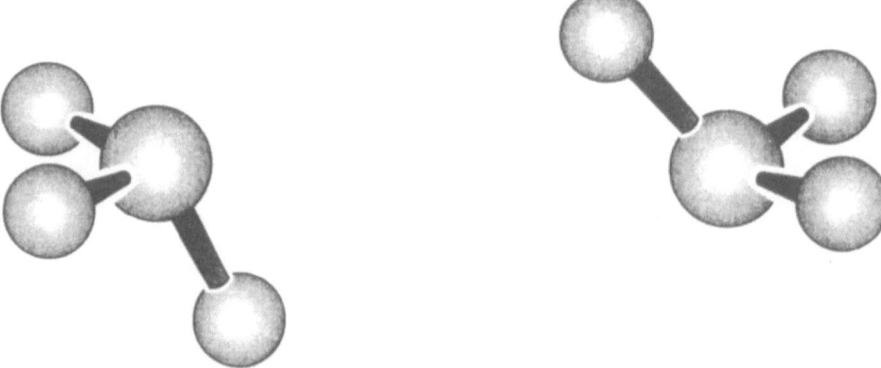

Fig. 9.4. Structure of $(NH_3)_2$. Displayed is the so-called cyclic structure as deduced from the microwave measurements of Klemperer and coworkers

---

[1] 1 D(ebye) $= 3.3358 \times 10^{-30}$ C m.

This non-hydrogen-bonded structure was surprising in view of the earlier assumptions. The question arose, therefore, whether dynamical effects might be so dominant, even in the asymmetric form, that tunneling converted the above nearly linear hydrogen-bonded structure into an average structure that was very different. The following tunneling motions between equivalent configurations may be possible in $(NH_3)_2$:

- The C_3 rotation. Each NH_3 monomer can rotate about its own symmetry axis. Each rotation is described by a variation of ϕ. The rotation takes place in a potential with three equivalent minima and gives rise to a specific tunneling splitting. The general characteristics of tunneling splittings in such a three-well potential are well known and are described in, for example, [60].
- The interchange tunneling (I). This is a tunneling motion by which the two monomers change their relative orientations. As a result of this tunneling the two monomers interchange their polar angles θ_1 and θ_2 and adopt a symmetrically equivalent position in between. This motion takes place in a double-well potential and gives rise to an observable tunneling splitting.
- The inversion tunneling. This motion corresponds to the well-studied umbrella inversion in the NH_3 monomer [82]. Initially, this motion was thought to be quenched completely, owing to the presence of the second ammonia molecule.

A schematic energy diagram of the expected tunneling splittings is shown in Fig. 9.5. A similar diagram is displayed in [141]. On the left side, the interchange tunneling is assumed to be small compared with the splitting due to the C_3 rotation; on the right side, the opposite behavior is assumed. The inversion tunneling splitting is neglected.

In general, one can state that the higher the barrier for a specific tunneling motion, the smaller is the corresponding tunneling splitting. If the barriers for all tunneling motions are sufficiently high we can consider $(NH_3)_2$ as a rigid molecule and can easily deduce the structure from the microwave measurements. If, however, the complex exhibits large-amplitude motions between symmetrically equivalent configurations, i.e. if it is not (nearly) rigid, the deduction of structural parameters from the measured quantities is no longer straightforward, as has been pointed out already for the case of Ar–CO. In this case the measured quantities are averaged over the corresponding tunneling paths. They therefore sample extensive regions of the potential surface, which implies that the measured vibrationally averaged structure may be quite different from the equilibrium structure.

The fact that isotopic substitution resulted in only small variations of the polar angles was taken as evidence that dynamical effects could be neglected in the structural determination. For a rigid hydrogen-bonded structure, a dipole moment of about 2 D is expected, whereas the measured value for $(NH_3)_2$ is 0.74 D. The dipole moment of the deuterated species $(ND_3)_2$ was found to be even smaller than for $(NH_3)_2$ [142]. Since the deuterated species

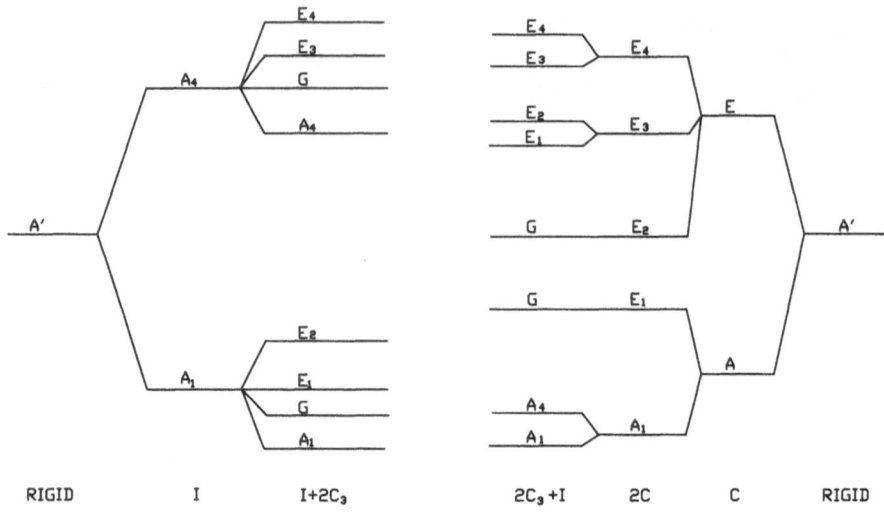

Fig. 9.5. Schematic energy-level scheme for $(NH_3)_2$. The two tunneling motions, interchange tunneling (I) and the C_3 tunneling motion (C_3), are assumed to be switched on in sequence. On the *left* $I < C_3$ and on the *right* $C_3 < I$ is assumed. The quantum numbers (j_1, j_2) correspond to the angular momentum of the internal rotation of the two NH_3 monomers

is heavier, its internal motions should be more hindered. Consequently, it should remain closer to equilibrium, which supports the assumption of a non-hydrogen-bonded equilibrium structure.

However, later ab initio studies showed a very flat potential surface [71, 185], which indicates the presence of large-amplitude motions, in contradiction to the previous conclusions, and calls into question the deduction of the cyclic structure. The barrier height in the potential surface between two symmetrically equivalent configurations can be deduced from the tunneling splittings between the corresponding energy levels. The measurement of these tunneling splittings is therefore a sensitive test of the potential surface.

9.4 The Far-Infrared Spectrum

Whereas transitions between different rotational levels are measured by use of microwave spectroscopy and vibrational transitions are probed by infrared (IR) spectroscopy, transitions between different tunneling levels lie in the far-infrared (FIR) region. The first measurement of the FIR spectrum of $(NH_3)_2$ was reported in 1990 by Havenith et al. [72] and was performed with the tunable far-infrared spectrometer at Berkeley. The initial assignment of the data was not in agreement with later FIR measurements from the Nijmegen group [203] (see also [73]). Therefore an IR–FIR double-resonance experiment was performed [73], which yielded a new assignment of the previously studied

IR and FIR transitions. At the same time a very extensive FIR study was performed by Loeser et al. [115] at Berkeley, who assigned all the transitions involved and obtained a complete set of energy levels and splittings.

Dynamical calculations were performed by van Bladel et al. [192]. Starting from a model potential based upon ab initio calculations [164], they explored the full six-dimensional vibration–rotation–tunneling dynamics. These new measurements and the dynamical calculations performed at the same time all came to the following conclusions: the small tunneling splitting which was observed previously [141] is not due to the interchange tunneling, but is due to the umbrella inversion of the monomers in the ground state. Earlier, this inversion was believed to be quenched completely, but in fact it appears to be only hindered and not totally quenched. The interchange tunneling splitting turned out to be of the order of 15–20 cm^{-1}. This large splitting indicates that $(NH_3)_2$ exhibits very-large-amplitude motions in the interchange coordinate. This conclusion gave rise to new speculations regarding the structure of $(NH_3)_2$ and initiated new experiments and calculations on this complex, which we describe in the next section.

9.5 The Infrared Spectrum

Between 1985 and 1988, IR studies utilizing the high power of the CO_2 laser, which has an overlap with the NH_3 umbrella vibration within the complex, were reported by several groups [76, 87, 179]. Fraser et al. [50] showed by infrared–microwave double resonance that the two states observed in the microwave spectrum had quite different infrared spectra. The two measured levels were attributed to the inequivalent ortho–para and para–ortho states of the ammonia dimer. These states have no interchange tunneling to first order and it was inferred from the measurements that interchange tunneling was quenched in these states [142, 143]. This seemed to reinforce the conclusion that the equilibrium structure should be close to the cyclic structure.

The studies of Snels et al. [179] yielded the transition frequencies of two bands which were attributed to the donor and acceptor ν_2 excitation. Together with measurements by Bacic et al. [6], the dissociation energy of the dimer was determined to lie between 6 and 12 kJ/mol (500–1000 cm^{-1}). From ab initio calculations D_e was calculated to be 1043 cm^{-1}[185]. If we subtract the zero-point energy of the six soft modes (about 385 cm^{-1}), which was calculated later [146], we obtain a dissociation energy D_0 of 658 cm^{-1}. The results of the FIR–IR double-resonance study by Havenith et al. [73] yielded a new assignment of the previous IR transitions. A splitting of 4 cm^{-1} at 980 cm^{-1} could be assigned as the hindered inversion splitting in the excited state. A complete assignment of the IR spectra is given in a later paper by Linnartz et al. [112]. Following the argument of Fraser [51] for $(HF)_2$, a qualitative model was introduced which explained the experimental results. In the excited state two different potential curves exist. Each potential curve is asymmetric owing to the difference in excitation energy for the two nonequivalent monomers. However, these two curves are coupled by an interaction term V_{ab}, which is proportional to μ_{01}^2/R^3, where μ_{01} is the transition dipole moment and R the distance between the two monomers. As a result two symmetric potential curves are obtained, in which the barrier is changed. In the following, the two IR bands which correspond to transitions to bound states of these two potentials are designated as ν_2 (976 cm^{-1}) and ν_1 (1004 cm^{-1}). The authors concluded that in the ν_2 band the interchange motion is hindered, such that the corresponding splitting is reduced from 20 cm^{-1} in the ground state to less than 1–2 cm^{-1}.

9.6 Dipole and Quadrupole Measurements

If large-amplitude motions play an important role, measurements will yield quantities averaged over broad regions of the intermolecular potential surface. Different quantum states probe different parts of the potential surface and therefore the averaged values will depend on the specific quantum state probed. The structure of $(NH_3)_2$ shown in Fig. 9.4 was deduced from the measurement of the dipole moment and nuclear quadrupole splittings in *one* specific quantum state. In the state which was probed by Nelson et al. [140] the dimer consists of one ortho (total proton spin $I = 3/2$) and one para ($I = 1/2$) ammonia molecule. The state is characterized further by a vanishing projection of the total angular momentum onto the a axis of the complex, i.e. a quantum number $K = 0$. A good test of the existence of large-amplitude motions is therefore provided by the measurement of these quantities in other quantum states.

The electric dipole moments of two ortho–para ammonia dimer states with $K = 1$ were measured [111, 38] to be 0.10 D and 0.76 D. If the ammonia dimer is rigid the dipole moment is expected to be the same for all states. This is obviously not the case, which proves that dynamical effects are important. The presence of such large dynamical effects implies that a direct derivation of the equilibrium structure from the measured spectroscopic quantities is not possible.

The quadrupole splittings in a state consisting of two equivalent para NH_3 molecules were determined from measurements on the pulsed molecular-beam Fourier transform spectrometer in Kiel [77]. In these states tunneling must occur and motional averaging of the hyperfine operators will take place. These measurements probed for the first time the dihedral angle γ. The observed hyperfine pattern could not be predicted using the quadrupole coupling constants obtained by Nelson et al. [140]. The analysis and interpretation of the new data required a derivation [77] of the nuclear quadrupole splitting from the vibration–rotation–tunneling wave functions of $(NH_3)_2$ [146] which takes into account the dynamical character of the problem.

9.7 The Potential for (NH$_3$)$_2$

At the same time, more progress came from theory: Olthof et al. [146] constructed a new model potential for (NH$_3$)$_2$. Some parameters in this potential were adjusted to reproduce the observed FIR spectrum [115]. With the improved potential, the calculated energy levels of the different vibrational–rotational–tunneling states agree to within about 1% accuracy for all the states investigated experimentally. Also, the measured dipole moment and quadrupole splittings are reproduced quite well, giving an independent measure of the quality of the calculated wave functions. The use of these wave functions in perturbational calculations [147] of the small tunneling splittings caused by the hindered inversion of the NH$_3$ monomers was found to explain all the measured splittings [115], not only qualitatively but even quantitatively. The equilibrium structure obtained corresponds neither to the nearly linear hydrogen-bonded structure expected by comparison with other hydrogen-bonded systems such as (H$_2$O)$_2$ and (HF)$_2$ nor to the cyclic structure (see Fig. 9.6). The structure is instead strongly bent, with $\theta_1 = 40°$ and $\theta_2 = 85°$, whereas for the linear hydrogen-bonded structure these angles would be $\theta_1 = 0°$ and $\theta_2 = 112°$.

The structure is actually nonrigid, and the barrier for interchange tunneling is very small (7 cm^{-1}) in comparison with the binding energy (1000 cm^{-1}). That this barrier is indeed very small was confirmed experimentally by new IR–FIR double-resonance measurements [112]. This means that a broad range of polar angles are accessed, which is in contradiction to the idea of a strong directionality as normally assumed to be typical for hydrogen bonding. However, the attractive electrostatic interaction between the protons and the lone pairs of the NH$_3$ monomers allows only *coupled* internal motions of the two monomers along the interchange tunneling path, which implies that directionality is not completely lost. In summary, the ammonia dimer is a highly nonrigid system and all measured quantities are affected substantially by vibrational averaging.

Fig. 9.6. Equilibrium structure as calculated by van der Avoird and coworkers

In addition, the improved potential was used to study the effect of isotopic substitution, or, more precisely, the effect of deuteration. The result was that the observed small changes of the measured quantities such as the dipole moment, which normally indicate the absence of vibrational averaging effects, could be well reproduced with this highly nonrigid model. It turned out that this effect, which misled the experimentalists in the beginning, is a consequence of the different hindered-rotor behaviors of the ortho and para monomers in the system that was initially studied by Nelson et al. [140].

The bonding in $(NH_3)_2$ is strong: $D_c = 1020$ cm^{-1} [146], which is about 65% of the binding energy of the hydrogen-bonded $(H_2O)_2$ and $(HF)_2$ complexes. The binding is still dominated by electrostatics, but one finds an increase in the relative importance of the less directional dispersion forces in progressing from $(HF)_2$ and $(H_2O)_2$ to $(NH_3)_2$ [28]. Moreover, the electron density localized near the hydrogen atoms increases significantly in the same order. Therefore, the exchange repulsion increases. It is this effect, the repulsion between the electron density at the hydrogen atoms and the lone pair of the nitrogen atoms, which was previously underestimated. This effect prevents the linear hydrogen-bonded structure and accounts for the bent equilibrium structure of $(NH_3)_2$. Hence, the use of simple electrostatic arguments to describe the anisotropy of the intermolecular potential fails and one must explicitly consider the effect that the exchange repulsion has on the geometry. The bonding in $(NH_3)_2$ has turned out to be more subtle than initially thought. This teaches us that the conventional notions and models regarding hydrogen bonding have to be improved in the future.

9.8 $(NH_3)_2$ in Helium Clusters

Measurements of the IR spectrum of an ammonia dimer embedded in a cold helium cluster have been performed very recently. A detailed description will be given in a forthcoming paper [7]. The measurements were performed using a line-tunable CO_2 laser. The helium cluster beam was generated by expanding pure helium through a 5 μm diameter nozzle at a temperature of $T_0 = 20$ K and a stagnation pressure of $p_0 = 40$ bar, yielding a mean cluster size of about 4000. These clusters entered a scattering chamber containing ammonia gas, where they could capture ammonia molecules, which then formed a dimer within the helium cluster. This technique, the so-called pickup technique, was pioneered by Gough et al. [63] and was used in subsequent high-resolution studies of molecules embedded in large helium clusters, e.g. [70]. In correspondence with the previous data, the ammonia dimer was found to be in thermal equilibrium with the helium cluster, which means at temperatures of ca. 0.4 K.

The following tentative conclusions could be drawn from the experimental data:

- Inversion tunneling is not influenced by the presence of helium: the IR transitions show no substantial shift due to helium. The inversion tunneling splittings are equal to the inversion tunneling splittings for the free $(NH_3)_2$ complex (4–5 cm^{-1}). This contrasts with the situation in the NH_3 monomer, where a large blue shift was found to result from the presence of helium. The blue shift can be understood as a cage effect. Being a very-large-amplitude motion, the inversion tunneling is hindered by the helium; the potential is more repulsive at the walls and the frequency is blue-shifted. An even more dramatic effect occurs if we go from the NH_3 monomer to $(NH_3)_2$. In this case, one NH_3 monomer hinders the inversion tunneling motion of the second NH_3, the inversion tunneling splitting is decreased by a factor of 10 and the IR umbrella mode is blue-shifted by ca. 40 cm^{-1}. This implies that ammonia itself suppresses the inversion tunneling more effectively than the helium, which explains qualitatively why the further addition of helium does not affect the inversion tunneling any further.

- The situation for the interchange tunneling is quite different. The interchange motion corresponds to a large-amplitude motion in the free $(NH_3)_2$ complex. The interchange tunneling is strongly affected by the presence of helium atoms, which cause an increase of the tunneling barrier for this specific motion. The tunneling barrier for the interchange tunneling in the free $(NH_3)_2$ complex was found to be extremely small (7 cm^{-1}), which means that even small changes of the tunneling barrier could cause a noticeable change of the energy pattern. Experimentally, we found that the interchange tunneling in the ν_1 band was reduced from 5–10 cm^{-1} to less than 2 cm^{-1}. This implies that the interchange tunneling motion is hindered or quenched by the helium.

- The structure of the $(NH_3)_2$ complex has been the subject of a long debate. Since it was found that $(NH_3)_2$ exhibits large-amplitude motions, the discussion has focused on the question of whether we still have a hydrogen-bonded equilibrium structure or a more cyclic structure. The equilibrium structure obtained from the potential of van der Avoird and coworkers is shown in Fig. 9.6. It corresponds more to a cyclic ($\theta_1 = \theta_2$) than to a hydrogen-bonded ($\theta_1 = 0°$, $\theta_2 = 112°$) structure. The barrier for interchange tunneling is very small, which means that the position of the minimum is not exactly known. In this context the spectra of $(NH_3)_2$ in helium are especially interesting. The barriers for interchange tunneling are increased; the interchange motion is quenched considerably compared with the gas phase. This implies that any structural measurement of $(NH_3)_2$ in helium reflects the equilibrium structure much better than any measurement of the free $(NH_3)_2$ complex, where the vibrational averaging effects are by no means negligible. We observe that the difference between the ν_1 and ν_2 bands is considerably smaller for $(NH_3)_2$ in helium (14 cm^{-1}) than for free $(NH_3)_2$ (25 cm^{-1}), but much more than the splitting expected from a resonant dipole–dipole interaction. This might indicate that the two ammonia monomers are even more closely equivalent in the presence of helium. The near-identical positions of the two ammonia monomers, as were found in the study of Nelson et al., can therefore not be attributed exclusively to vibrational averaging. In the case of vibrational averaging between two distinct equilibrium structures, the hindering of the interchange tunneling motion should result in an increase of the difference between the ν_1 and ν_2 bands ($\nu_1 - \nu_2$). Instead, a decrease was found experimentally for $(NH_3)_2$ in a helium cluster. We can therefore state that our experimental observation supports the assumption of a more cyclic equilibrium structure for the $(NH_3)_2$ complex in helium.

For these reasons the ammonia dimer in helium should be an ideal candidate for studying the structure of the complex without the influence of the large-amplitude tunneling motions. However, a more definite answer to this problem would require high-resolution studies in the FIR and microwave regions, which are a challenge for the future.

10. Final Remarks

The point of this book was to give an overview of the issues that arise in considering the spectroscopy of intermolecular complexes, using a few examples to illustrate the important points. An introduction to the theory of intermolecular potentials was given in Chaps. 2 and 3.

Chapters 4–7 presented some experimental techniques which are currently used for the study of weakly bound complexes. In Chap. 4 the basic concepts of molecular beams were explained. Chapter 5 gave an overview of the infrared spectroscopy of molecular complexes. The following two chapters focused on the two spectrometers which were set up in Bonn during my Habilitation. We combined a diode laser spectrometer with a supersonic expansion, which yielded new spectra of the Ar–CO complex. These spectra provided new information on the intermolecular potential of this prototype system. The new data enabled us, with the help of G. Jansen from the theoretical chemistry department, to obtain a new, accurate potential for this complex, which was the subject of Chap. 8. The second spectrometer was a CO sideband laser spectrometer, which was combined with optothermal detection. The laser was optimized and a new kind of laser source could be established for the spectroscopy of molecular beams. This allowed, for the first time, the spectroscopy of complexes in the chemically important carbonyl region. A detailed description of the setup was given in Chap. 7.

Chapter 9 describes the results of the measurements on the ammonia dimer, which turned out to be a very important and interesting complex. From these studies, which were carried out in several laboratories, it became clear that some of the concepts which are normally used in molecular spectroscopy, such as structure and hydrogen bonding, might fail, or are at least a little too naive.

Considering the two prototype systems Ar–CO and $(NH_3)_2$, we could show the importance of dynamical effects in these complexes. A meaningful interpretation of the experimental data therefore required high-quality theoretical studies. However, the two examples demonstrate that pure ab initio calculations do not yet reach the required accuracy for a determination of the intermolecular potential. New methods are being developed and have profited from the existence of high-resolution studies. We can see that we can use these small systems to test and improve simple models. These models could then serve to describe more complicated systems which are of major importance in biology, such as peptides and proteins. It is a challenge for the future to perform new experimental and theoretical studies of these more complex systems.

References

1. R. Ahlrich, R. Penco, G. Scoles, Chem. Phys. **90**, 2181 (1989).
2. D.T. Anderson, S. Davis, T.S. Zvier, D.J. Nesbitt, Chem. Phys. Lett. **258**, 207 (1996).
3. J.G. Ángyán, G. Jansen, M. Loos, C. Hättig, B.A. Heß, Chem. Phys. Lett. **219**, 267 (1994).
4. E. Bachem, A. Dax, T. Fink, A. Weidenfeller, M. Schneider, W. Urban, Appl. Phys. B **57**, 185 (1993).
5. G. Bacher, N. Mais, M. Illing, A. Forchel, K. Schull, J. Nurnberger, G. Landvehr, Electron. Lett. **31**, 2184 (1995).
6. Z. Bacic, U. Buck, H. Meyer, R. Schinke, Chem. Phys. Lett. **125**, 47 (1986).
7. M. Behrens, U. Buck, R. Fröchtenicht, M. Hartmann, M. Havenith, J. Chem. Phys. **107**, 7179 (1997).
8. R.J. Bemish, P.A. Block, L.G. Pedersen, W. Yang, R.E. Miller, J. Chem. Phys. **99**, 8585 (1993).
9. P.F. Bernath, *Spectra of Atoms and Molecules*, Oxford University Press, Oxford (1995).
10. M.L. Bhaumik, Appl. Phys. Lett. **17**, 188 (1970).
11. D.D. Bicanic, B.F.J. Zuidberg, A. Dymanus, Appl. Phys. Lett. **32**, 367 (1978).
12. G. Bisonette, C.E. Chuaqui, K.G. Crowell, R.J. Le-Roy, R.J. Wheatley, W.J. Meath, J. Chem. Phys. **105**, 2639 (1996).
13. G.A. Blake, K.B. Laughlin, R.C. Cohen, K.L. Busarow, D.H. Gwo, C.A. Schmuttenmaer, D.W. Steyert, R.J. Saykally, Rev. Sci. Instrum. **62**, 1701 (1991).
14. E. Bonek, M. Knecht, G. Magerl, K. Peis, K.R. Richter, Arch. Elek. Uebertragung **32**, 209 (1978).
15. G. Brocks, A. van der Avoird, B.T. Sutcliffe, J. Tennyson, Mol. Phys. **50**, 1025 (1983).
16. U. Buck, H. Meyer, Phys. Rev. Lett. **52**, 109 (1984).
17. U. Buck, H. Meyer, J. Chem. Phys. **84**, 4854 (1986).
18. U. Buck, Adv. Atom. Mol. Opt. Phys. **35**, 121 (1995).
19. A.D. Buckingham, P.W. Fowler, J. Chem. Phys. **79**, 6426 (1983).
20. A.D. Buckingham, P.W. Fowler, Can. J. Chem. **63**, 2018 (1985).
21. A.D. Buckingham, P.W. Fowler, A.J. Stone, Internat. Rev. Phys. Chem. **5**, 107 (1986).
22. K.L. Busarow, G.A. Blake, K.B. Laughlin, R.C. Cohen, Y.T. Lee, R.J. Saykally, Chem. Phys. Lett. **141**, 289 (1987).
23. E.J. Buske, S.A. Nizkovodov, F.R. Bennet, J.P. Maur, J. Chem. Phys. **102**, 5152 (1995).
24. S.W. Bustamente, M. Okumura, D. Gerlich, H.S. Kwok, L.R. Carlson, Y.T. Lee, J. Chem. Phys. **86**, 508 (1987).
25. M.P. Casassa, D.S. Bomse, K.C. Janda, J. Chem. Phys. **74**, 5044 (1981).

26. V. Castells, N. Halberstadt, S.K. Shin, R.A. Beaudet, C. Wittig, J. Chem. Phys. **101**, 1006 (1994).
27. M. Cavallini, L. Meneghetti, G. Scoles, M. Yealland, Rev. Sci. Instrum. **42**, 1759 (1971).
28. G. Chalasiniski, M.M. Szczesniak, Chem. Rev. **94**, 1723 (1994).
29. M.-C. Chan, A.R.W. McKellar, J. Chem. Phys. **105**, 7910 (1996).
30. P.K. Cheon, IEEE J. Quant. Electr. **20**, 700 (1984).
31. J.M. Chevalier, J. Legrand, P. Gloriux, J. Chem. Phys. **90**, 6833 (1989).
32. S.E. Choi, J.C. Light, J. Chem. Phys. **97**, 7031 (1992).
33. C.E. Chuaqui, R.J. Le Roy, A.R.W. McKellar, J. Chem. Phys. **101**, 39 (1994).
34. V.J. Cocoran, J.M. Martin, W.T. Smith, Appl. Phys. Lett. **22**, 517 (1973).
35. R.C. Cohen, K.L. Busarow, C.A. Schmuttenmaer, Y.T. Lee, R.J. Saykally, Chem. Phys. Lett. **164**, 321 (1989).
36. R.C. Cohen, R.J. Saykally, J. Chem. Phys. **98**, 6007 (1993).
37. K.R. Comer, S.C. Foster, Chem. Phys. Lett. **164**, 321 (1994).
38. G. Cotti, H. Linnartz, W.L. Meerts, A. van der Avoird, E.H.T. Olthof, J. Chem. Phys. **104**, 3898 (1996).
39. R.F. Curl, K.K. Murray, M. Petri, M.L. Richnow, F.K. Tille, Chem. Phys. Lett. **161**, 98 (1989).
40. R.L. DeLeon, J.S. Muenter, J. Chem. Phys. **80**, 6092 (1984).
41. W. Demtröder, *Laser Spectroscopy*, Vol. V, 2nd ed., Springer, Berlin, Heidelberg (1996).
42. A. DePiante, E.J. Campbell, S.J. Buelow, Rev. Sci. Instrum. **60**, 858 (1989).
43. C. Douketis, G. Scoles, S. Marchetti, M. Zen, A.J. Thakkar, J. Chem. Phys. **76**, 3057 (1982).
44. C. Douketis, J.M. Hutson, B.J. Orr, G. Scoles, Mol. Phys. **52**, 763 (1984).
45. A.T. Droege and P.C. Engelking, Chem. Phys. Lett. **96**, 316 (1983).
46. T.R. Dyke, B.J. Howard, W. Klemperer, J. Chem. Phys. **56**, 2442 (1972).
47. M.J. Elrod, R.J. Saykally, J. Chem. Phys. **103**, 933 (1995).
48. R.S. Eng, J.F. Butler, K.J. Linden, Opt. Eng. **19**, 945 (1980).
49. P. Engels, Diplom thesis, University of Bonn (1996).
50. G.T. Fraser, D.D. Nelson Jr., A. Charo, W. Klemperer, J. Chem. Phys. **82**, 2535 (1985).
51. G.T. Fraser, J. Chem. Phys. **90**, 2097 (1989).
52. G.T. Fraser, A.S. Pine, J. Chem. Phys. **91**, 637 (1989).
53. G.T. Fraser, A.S. Pine, W.A. Kreiner, J. Chem. Phys. **94**, 7061 (1991).
54. C. Freed, Appl. Phys. Lett. **18**, 458 (1971).
55. M.J. Frisch, J.E. Del Bene, J.S. Binkley, H.F. Schaefer, J. Chem. Phys. **84**, 2279 (1986).
56. J.E. Gambogi, M. Becucci, C.J. O'Brian, K.K. Lehmann, G. Scoles, Ber. Bunsenges. Phys. Chem. **99**, 548 (1995).
57. T. George, B. Wu, A. Dax, M. Schneider, W. Urban, Appl. Phys. B **59**, 159 (1994).
58. F.A. Gianturco, F. Paesani, J. Chem. Phys. **115**, 249 (2001).
59. G.M. Gibsen, M. Ebrahimzadeh, M.S. Padgetl, M.H. Dunn, Optics Lett. **24**, 397 (1999).
60. W. Gordy, R.L. Cook, *Microwave Molecular Spectra*, Interscience, Chichester (1970).
61. T.E. Gough, R.E. Miller, G. Scoles, Appl. Phys. Lett. **30**, 338 (1977).
62. T.E. Gough, R.E. Miller, G. Scoles, J. Chem. Phys. **69**, 1588 (1978).
63. S. Goyal, D.L. Schutt, G. Scoles, Phys. Rev. Lett. **69**, 933 (1992).
64. M. Gromoll-Bohle, W. Bohle, W. Urban, Opt. Commun. **69**, 409 (1989).
65. T. Häber, U. Schmitt, M. Suhm, Phys. Chem. Chem. Phys. **1**, 5573 (1999).

66. Ch. Hättig, B.A. Heß, J. Chem. Phys. **105**, 9948 (1996).
67. Ch. Hättig, B.A. Heß, J. Phys. Chem. **100**, 6243 (1996).
68. O. Hagena, Surf. Sci., **106**, 101 (1981).
69. K. Harada, T. Tanaka, Chem. Phys. Lett. **227**, 65 (1994).
70. M. Hartmann, R.E. Miller, J.P. Toennies, A.F. Vilesov, Phys. Rev. Lett. **75**, 1566 (1995).
71. D.M. Hassett, C.J. Marsden, B.J. Smith, Chem. Phys. Lett. **183**, 449 (1991).
72. M. Havenith, R.C. Cohen, K.L. Busarow, D.H. Gwo, Y.T. Lee, R.J. Saykally, J. Chem. Phys. **94**, 4776 (1991).
73. M. Havenith, H. Linnartz, E. Zwart, A. Kips, J.J. ter Meulen, W.L. Meerts, Chem. Phys. Lett. **193**, 261 (1992).
74. M. Havenith, G. Hilpert, M. Petri, W. Urban, Mol. Phys. **81**, 1003 (1994).
75. G.D. Hayman, J. Hodge, B.J. Howard, J.S. Muenter, T.R. Dyke, Chem. Phys. Lett. **118**, 12 (1985).
76. B. Heijmen, A. Bizzari, S. Stolte, J. Reuss, Chem. Phys. **126**, 201 (1988).
77. N. Heineking, W. Stahl, E.H.T. Olthof, P.E.S. Wormer, A. van der Avoird, M. Havenith, J. Chem. Phys. **102**, 8693 (1995).
78. J. Hepburn, G. Scoles, R. Penco, Chem. Phys. Lett. **36**, 451 (1975).
79. M. Hepp, W. Jäger, I. Pak, G. Winnewisser, J. Mol. Spec. **176**, 58 (1996).
80. D. Herriott, H. Kogelnik, R. Kompfer, Appl. Opt. **3**, 523 (1964).
81. D. Herriott, H. Schulte, Appl. Opt. **4**, 883 (1965).
82. G. Herzberg, *Molecular Spectra and Molecular Structure*, Vol. II: *Infrared and Raman Spectra of Polyatomic Molecules*, Van Nostrand, Princeton (1945).
83. G. Hilpert, H. Linnartz, M. Havenith, J.J. ter Meulen, W.L. Meerts, Chem. Phys. Lett. **219**, 384 (1994).
84. M.A. Hoffbauer, K. Liu, C.F. Giese, W.R. Gentry, J. Chem. Phys. **78**, 5567 (1983).
85. J.M. Hollas, *High Resolution Spectroscopy*, Wiley, Chichester (1998).
86. S.C. Hsu, R.H. Schwendemann, G. Magerl, IEEE J. Quant. Electr. **24**, 2294 (1988).
87. F. Huisken, T. Pertsch, Chem. Phys. **126**, 213 (1988).
88. F. Huisken, M. Kaloudis, A. Kulcke, C. Laush, J.M. Lisy, J. Chem. Phys. **103**, 5366 (1995).
89. F. Huisken, O. Werhahn, A Yu Ivanor, S.A. Krasnokutski, J. Chem. Phys. **111**, 2978 (1999).
90. J.M. Hutson, J. Chem. Phys. **89**, 4550 (1988).
91. J.M. Hutson, J. Chem. Phys. **91**, 4455 (1989).
92. J.M. Hutson, Ann. Rev. Phys. Chem. **41**, 123 (1990).
93. J.M. Hutson, J. Chem. Phys. **96**, 6752 (1992).
94. J.M. Hutson, J. Chem. Phys. **105**, 9130 (1996).
95. W. Jäger, M.C.L. Gerry, J. Chem. Phys. **102**, 3587 (1995).
96. G. Jansen, J. Chem. Phys. **105**, 89 (1996).
97. E. Kapon, A. Katzir, IEEE J. Quant. Electron. **21**, 1947 (1985).
98. D.A. Kirkvood, H. Linnartz, M. Grutter, O. Dopfer, T. Motylevski, M. Pakkov, M. Tulej, M. Wyss, J.P. Maier, Faraday Discuss **109**, 109 (1998).
99. S. König, G. Hilpert, M. Havenith, Mol. Phys. **86**, 1233 (1995).
100. S. König, M. Havenith, Mol. Phys. **91**, 265 (1997).
101. P.P. Korambath, X.T. Wu, E.F. Hayes, J. Phys. Chem. **100**, 6116 (1996).
102. F. Kühnemann, K. Schneider, A. Hecker, A.A.E. Martis, W. Urban, S. Schiller, J. Mlynek, Appl. Phys. **66**, 741 (1998).
103. B. Kukawska-Tarnawska, G. Chalasinski, K. Olszewski, J. Chem. Phys. **101**, 4964 (1994).
104. A. Kumar, W.J. Meath, Mol. Phys. **54**, 823 (1985).

105. W. Kutzelnigg, *Einführung in die Theoretische Chemie*, Vol. 2, Chap. 15, Verlag Chemie, Weinheim (1978).
106. Z. Latajka, S. Schreiner, J. Chem. Phys. **84**, 341 (1986); and references therein.
107. R.J. Le-Roy, J.M. Hutson, J. Chem. Phys. **86**, 837 (1987).
108. K.R. Leopold, G.T. Fraser, S.E. Novick, W. Klemperer, Chem. Rev. **94**, 1807 (1994).
109. K.J. Linden, A.W. Mantz, SPIE Proc. **320**, 109 (1982).
110. K.J. Linden, IEEE J. Quant. Electron. **21**, 391 (1985).
111. H. Linnartz, A. Kips, W.L. Meerts, M. Havenith, J. Chem. Phys. **99**, 2449 (1993).
112. H. Linnartz, W.L. Meerts, M. Havenith, Chem. Phys. **193**, 327 (1995).
113. H. Linnartz, D. Verdes, Th. Speck, Rev. of Scientific Instruments **71**, 1811 (2000).
114. S. Liu, C.E. Dykstra, K. Kolenbrander, J.M. Lisy, J. Chem. Phys. **85**, 2077 (1986).
115. J.G. Loeser, C.A. Schmuttenmaer, R.C. Cohen, M.J. Elrod, D.W. Steyert, R.J. Saykally, R.E. Bumgarner, G.A. Blake, J. Chem. Phys. **97**, 4727 (1992).
116. C.M. Lovejoy, D.J. Nesbitt, Rev. Sci. Instrum. **58**, 807 (1987).
117. C.M. Lovejoy, D.J. Nesbitt, J. Chem. Phys. **86**, 3151 (1987); J. Chem. Phys. **60**, 858 (1989).
118. C.M. Lovejoy, D.J. Nesbitt, Chem. Phys. Lett. **146**, 582 (1988).
119. C.M. Lovejoy, J.M. Hutson, D.J. Nesbitt, J. Chem. Phys. **97**, 8009 (1992).
120. D. Luckhaus, M. Quack, U. Schmitt, M. A. Suhm, Ber. Bunsenges. Phys. Chem. **99**, 457 (1995).
121. G. Magerl, E. Bonek, W.A. Kreiner, Chem. Phys. Lett. **52**, 473 (1977).
122. G. Magerl, E. Bonek, Appl. Phys. Lett. **34**, 452 (1979).
123. G. Magerl, W. Schupita, E. Bonek, W.A. Kreiner, J. Mol. Spec. **83**, 431 (1980).
124. G. Magerl, W. Schupita, E. Bonek, IEEE J. Quant. Electr. **18**, 1214 (1982).
125. G.C. Maitland, M. Rigby, E.B. Smith, W.A. Wakeham, *Intermolecular Forces: Their Origin and Determination*, Oxford University Press, Oxford (1981).
126. A.R.W. McKellar, Y.P. Zeng, S.W. Sharpe, C. Wittig, R.A. Beudet, J. Mol. Spec. **153**, 475 (1992).
127. R.F. Meads, A.L. McIntosh, J.I. Arno, C.L. Hartz, R.R. Luccheses, J. W. Bevan, J. Chem. Phys. **101**, 4593 (1994).
128. W.L. Meerts, F.H. de Leeuw, A. Dynamus, Chem. Phys. **22**, 314 (1977).
129. U. Merker, Diplom thesis, University of Bonn (1994).
130. U. Merker, P. Engels, F. Madeja, M. Havenith, W. Urban, Rev. Sci. Instrum. **70**, 1933 (1999).
131. B. Meyer, S. Saupe, M.H. Wappelhorst, T. George, F. Kühnemann, M. Schneider, M. Havenith, W. Urban, J. Legrand, Appl. Phys. B **61**, 169 (1995).
132. R.E. Miller, J. Phys. Chem. **90**, 3301 (1986).
133. R.E. Miller, in G. Scoles (ed.), *Atomic and Molecular Beam Methods*, Vol. II, Oxford University Press, Oxford (1992).
134. K. Mirsky, Chem. Phys. **46**, 445 (1980).
135. Y. Mizugai, H. Kuze, H. Jones, M. Takami, Appl. Phys. B **32**, 43 (1993).
136. R. Moszynski, T. Korona, P.E.S. Wormer, A. van der Avoird, J. Chem. Phys. **103**, 1 (1995).
137. J.M. Murrel, G. Shaw, J. Chem. Phys. **46**, 1768 (1967).
138. J.I. Musher, A.T. Amos, Phys. Rev. **164**, 31 (1967).
139. G. Nelke, Diplom thesis, University of Bonn (1992).
140. D.D. Nelson Jr., G.T. Fraser, W. Klemperer, J. Chem. Phys. **83**, 6201 (1985).
141. D.D. Nelson Jr., W. Klemperer, J. Chem. Phys. **87**, 139 (1987).

142. D.D. Nelson Jr., W. Klemperer, G.T. Fraser, F.J. Lovas, R.D. Suenram, J. Chem. Phys. **87**, 6365 (1987).

143. D.D. Nelson Jr., G.T. Fraser, W. Klemperer, Science **238**, 1670 (1988).

144. D.J. Nesbitt, Chem. Rev. **88**, 843 (1988).

145. T. Ogata, W. Jäger, I. Ozier, M.C.L. Gerry, J. Chem. Phys. **98**, 9399 (1993).

146. E.H.T. Olthof, A. van der Avoird, P.E.S. Wormer, J. Chem. Phys. **101**, 8430 (1994).

147. E.H.T. Olthof, A. van der Avoird, P.E.S. Wormer, J.G. Loeser, R.J. Saykally, J. Chem. Phys. **101**, 8443 (1994).

148. J. Orr, T. Oka, Appl. Phys. **21**, 293 (1980).

149. C.A. Parish, J.D. Augspurger, C.E. Dykstra, J. Phys. Chem. **96**, 2069 (1992).

150. C.K.N. Patel, Phys. Rev. **141**, 71 (1966).

151. L. Pauling, Proc. Nat. Acad. Sci. **14**, 359 (1928).

152. A.C. Peet, W.J. Yang, J. Chem. Phys. **91**, 6598 (1989).

153. M. Petri, Ph.D. thesis, University of Bonn 1991.

154. A.S. Pine, W.J. Lafferty, J. Chem. Phys. **78**, 2154 (1983).

155. A.S. Pine, G.T. Fraser, J. Chem. Phys. **89**, 100 (1988).

156. A.S. Pine, R.D. Suenram, E.R. Brown, K.A. McIntosh, J. Mol. Spec. **175**, 37 (1996).

157. D.G. Prichard, R.N. Nandi, J.S. Muenter, J. Chem. Phys. **89**, 115 (1988).

158. M. Quack, M. Suhm, J. Chem. Phys. **95**, 28 (1991).

159. A.E. Reed, L.A. Curtiss, F. Weinhold, Chem. Rev. **88**, 899 (1988).

160. S.W. Reeve, M.A. Dvorak, D.W. Firth, K.R. Leopold, Chem. Phys. Lett. **181**, 259 (1991).

161. S. Richter, M. Havenith, G. Jansen, to be published.

162. M. Rigby, E.B. Smith, W.A. Wakeham, G.C. Maitland, *The Forces Between Molecules*, Oxford University Press, Oxford (1986).

163. T. Ruchti, A. Rohrbacher, T. Speck, J.P. Connelly, E.J. Bieske, J.P. Maier, Chem. Phys. Lett. **209**, 169 (1996).

164. K.P. Sagarik, R. Ahlrichs, S. Bröde, J. Chem. Phys. **57**, 1247 (1986).

165. M. Schaefer, Ph.D. thesis, University of Bonn (1995).

166. I. Scheele, R. Lehnig, M. Havenith, Mol. Phys. **99**, 197 (2001).

167. I. Scheele, R. Lehnig, M. Havenith, Mol. Phys. **99**, 205 (2001).

168. J. Schleipen, J.J. ter Meulen, G.C.M. van der Sanden, P.E.S. Wormer, A. van der Avoird, Chem. Phys. **163**, 161 (1992).

169. C.A. Schmuttenmaer, R.C. Cohen, R.J. Saykally, J. Chem. Phys. **101**, 146 (1994).

170. H.J. Schneider, A. Yatsimirsky, *Principles and Methods in Supramolecular Chemistry*, John Wiley & Sons, New York (2000).

171. K. Schneider, P. Kramper, S. Schiller, J. Mlynek, Opt. Lett. **22**, 1293 (1997).

172. M. Schneider, A. Hinz, A. Groh, K.M. Evenson, W. Urban, Appl. Phys. B **44**, 241 (1987).

173. G. Scoles (ed.), *Atomic and Molecular Beam Methods*, Oxford University Press, Oxford (1988).

174. G. Scoles, M. Havenith, in preparation.

175. S.W. Sharpe, R. Sheeks, C. Wittig, R.A. Beaudet, Chem. Phys. Lett. **151**, 267 (1988).

176. S. Shin, S.K. Shin, F.-M. Tao, J. Chem. Phys. **104**, 183 (1996).

177. U. Simon, S. Waltman, I. Loa, F.K. Tittel, L. Hollberg, J. Opt. Soc. Am. B **12**, 323 (1995).

178. M. Snels, R. Fantoni, M. Zen, S. Stolte, J. Reuss, Chem. Phys. Lett. **124**, 1 (1986).

179. M. Snels, R. Fantoni, R. Sanders, W.L. Meerts, Chem. Phys. **115**, 79 (1987).

180. A.J. Stone, Chem. Phys. Lett. **83**, 233 (1981).
181. A.J. Stone, Chem. Phys. Lett. **211**, 101 (1993).
182. A.J. Stone, *The Theory of Intermolecular Forces*, Clarendon Press, Oxford (1996).
183. K.T. Tang, J.P. Toennies, J. Chem. Phys. **80**, 3726 (1984).
184. S. Taniguchi, T. Hino, S. Itho, K. Nakano, N. Nakayama, A. Ishibashi, M. Ikeda, Electron. Lett. **32**, 552 (1996).
185. F.M. Tao W. Klemperer, J. Chem. Phys. **99**, 5976 (1993).
186. J. Tennyson, S. Miller, B.T. Sutcliffe, J. Chem. Soc. Faraday Trans. II, **84**, 1295 (1988).
187. R.R. Toczylowski, S.M. Cybulski, J. Chem. Phys. **112**, 4604 (2000).
188. J.P. Toennies, Chem. Phys. Lett. **20**, 238 (1973).
189. C.E. Treanor, J.W. Rich, J. Chem. Phys. **48**, 1789 (1968).
190. W. Urban, in A.C.P. Alves, J.M. Brown, M. Hollas (eds.), *Frontiers of Laser Spectroscopy of Gases*, NATO ASI Series, Kluwer Academic, Dordrecht (1988).
191. W. Urban, Infrared Phys. Techn. **36**, 465 (1995).
192. J.W.I. van Bladel, A. van der Avoird, P.E.S. Wormer, J. Chem. Phys. **95**, 793 (1990).
193. A. van der Avoird, P.E.S. Wormer, R. Moszynski, Chem. Rev. **94**, 1931 (1994).
194. A. van der Avoird, G.T. Fraser, M. Havenith, W. Klemperer, H. Linnartz, J. Loeser, W.L. Meerts, E.H.T. Olthof, R.J. Saykally, W. Stahl, P.E.S. Wormer, Science, to be published.
195. P. Verhoeve, E. Zwart. M. Drabbels, J.J. ter Meulen, W.L. Meerts, A. Dymanus, D.B. McLay, Rev. Sci. Instrum. **61**, 1612 (1990).
196. M.F. Vernon, D.J. Krajnovich, H.S. Kwok, J.M. Lisy, Y.R. Shen, Y.T. Lee, J. Chem. Phys. **77**, 47 (1982).
197. F. Vögtle, *Supramolekulare Chemie*, Teubner, Stuttgart (1992).
198. D.J. Wales, A.J. Stone, P.L.A. Popelier, Chem. Phys. Lett. **240**, 89 (1995).
199. J.V. White, J. Opt. Soc. Am. **66**, 411 (1976).
200. P.E.S. Wormer, H. Hettema, J. Chem. Phys. **97**, 5592 (1992).
201. Y. Xu, M. Fukushima, T. Amano, A.R.W. McKellar, Chem. Phys. Lett. **242**, 126 (1995).
202. Y. Xu, A.R.W. McKellar, Mol. Phys. **88**, 859 (1996).
203. E. Zwart, Ph.D. thesis, University of Nijmegen (1991)

Index